# Vacuum Engineering Calculations, Formulas, and Solved Exercises

# Vacuum Engineering Calculations, Formulas, and Solved Exercises


Armand Berman

National Physical Laboratory of Israel
Hebrew University Givat Ram
Jerusalem



Academic Press, Inc.
*Harcourt Brace Jovanovich, Publishers*
San Diego New York Boston London Sydney Tokyo Toronto

This book is printed on acid-free paper. ∞



Academic Press, Inc.
1250 Sixth Avenue, San Diego, California 92101-4311

*United Kingdom Edition published by*
Academic Press Limited
24–28 Oval Road, London NW1 7DX

Armand Berman
P.O. Box 2243, Bat-Yam 59121
Israel

Library of Congress Cataloging-in-Publication Data

Berman, A. (Armand)
Vacuum engineering calculations, formulas, and solved exercises / Armand Berman.
p. cm.
Includes index.
ISBN 0-12-092455-2
1. Vacuum--Handbooks, Manuals, etc. 2. Vacuum--Problems, exercises, etc. 3. Vacuum technology--Handbooks, manuals, etc. 4. Vacuum technology--Problems, exercises, etc. I. Title.
QC166.B47 1992
621.5'5--dc20 92-4140
CIP

PRINTED IN THE UNITED STATES OF AMERICA
92 93 94 95 96 97 QW 9 8 7 6 5 4 3 2 1

# Contents

# Preface

Vacuum is now required for many scientific and industrial purposes, particularly those associated with semiconductors, lighting bulbs, and material-coating industries. The need to manufacture in highly characterized and controlled vacuum brings continual progress in the design, operation, and maintenance of vacuum systems, challenging the people involved in these activities. To many of these people, the use and application of vacuum engineering formulas rather than vacuum's theoretical basis are prime objectives.

This book is written for those who can devote little time for a full mathematical treatment of the many problems encountered in the vacuum practice but who have a reasonable knowledge of the essentials of vacuum together with elementary physics and mathematics. The text, which is intended to be a systematic answer to the manifold situations arising in practice, was written bearing in mind two main objectives: (1) to summarize and organize the vast material of the vacuum technology in sets of useful formulas, and (2) to frame a collection of worked-out exercises exemplifying how to use these formulas for solving technological problems.

The book starts with chapters on ideal (perfect) and real gases (Chapters 1 and 2) and the kinetic theory of gases (Chapter 3), which are intended to give the reader enough technique and experience with calculations.

Next, Chapter 4 deals with the flow rate of gases, throughput, and impedance of interconnected vacuum components.

There follow three chapters on the steady flow of gas in the viscous (Chapter 5), molecular (Chapter 6), and transition (Chapter 7) ranges, with emphasis on the viscous-chocked flow and molecular flow through short pipes, channels, and vacuum components of complex geometry. The extent of Chapter 7 on steady flow of gas in the transition range has been kept reduced, since for many practical vacuum systems it is sufficient to limit calculations to either steady viscous or molecular flow.

Chapter 8 on gas load examines the sources of gas in vacuum systems and evaluates the gas load. The treatment of gas adsorption and diffusion has been omitted, taking into account the aim of the book.

There then follows Chapter 9, devoted to vacuum pumps, and chapters dealing with the removal from a vacuum system of either the original gas under transient pressure (Chapter 10) or the residual gas under a pressure associated with the gas load divided by the available pumping speed (Chapter 11).

The derivation of the formulas is not given, since there are many authoritative textbooks where the basic theory is extensively discussed. For the reader interested in studying the subject in more detail, several references are listed at the end of each chapter.

All the equations are written in a form that permits the reader to use any system of units. In equations in which a combination of numerical quantities and symbols occurs, the units of measurement are stated. For the sake of clarity, the units of measurement have been written even in the intermediate steps of numerical calculations.

Although not universally adopted, the self-consistent system SI of units has been used throughout the text. To facilitate understanding for readers not yet accustomed to this system, other coherent and incoherent systems of units have been used. The Torr has been retained as a pressure unit, since it is very largely spread among vacuum workers.

Solved exercises, appended at the end of each chapter, were not chosen at random but rather have been carefully selected so as to include at least one representative example of each type of technological problem. Many exercises require equations from more than one chapter to permit greater flexibility in the coverage of those points having importance and usefulness in practice. The reader is strongly advised to check the accuracy of intermediary and final results of the exercises by using a dimensional equation.

The book is intended for the engineer, physicist, graduate student, or advanced technician engaged in purchasing, setting up, operating, or maintaining vacuum systems and equipment.

*Armand Berman*

# Acknowledgments

I gratefully acknowledge constant encouragement and helpful suggestions and criticism from the late Professor A. Roth. I should like to thank the authors, journals, and publishers cited in the text for permission to reproduce figures and tables. Many thanks are due to the staff of Academic Press for their kind help and cooperation in publishing this book. And to my wife, Ilane, who has tolerated preoccupations, shared disappointments, read many drafts, and still found nice words to say, I express my gratitude and devotion.

# List of Symbols

| | |
|---|---|
| $A$ | Area |
| $a$ | Linear dimension; constant |
| $a_{kg}$ | Amount of gas in mass units |
| $a_{m}$ | Amount of gas in molecule units |
| $a_{mol}$ | Amount of gas in mole units |
| $a_{pv}$ | Amount of gas in pressure–volume units |
| $B$ | Second virial coefficient |
| $\dot{B}$ | Volume of gas impinging on unit area per unit time |
| $B_{o}$ | Volume of gas escaping through an orifice per unit time |
| $b$ | Linear dimension; constant |
| $C$ | Conductance to gas flow |
| $C_{l}$ | Conductance to laminar flow |
| $C_{lch}$ | Conductance to chocked laminar flow |
| $C_{lnc}$ | Conductance to nonchocked laminar flow |
| $C_{m}$ | Conductance to molecular gas flow |
| $C_{ma}$ | Conductance to molecular gas flow of an aperture |
| $C_{md}$ | Conductance to molecular gas flow of a diaphragm |
| $C_{mp}$ | Conductance to molecular gas flow of a pipe, end effects disregarded |
| $C_{ms}$ | Conductance to molecular gas flow of rectangular slits |
| $C_{mT}$ | Conductance to molecular gas flow of a pipe, end effects accounted for |
| $C_{r}$ | Compression ratio of a mechanical pump |
| $C_{s}$ | Sutherland's constant |
| $C_{T}$ | Conductance of vacuum components connected in parallel |
| $C_{tr}$ | Conductance to transition flow |
| $C_{i}$ | Mole fraction of a gas component in a mixture of gases |
| $c_{p}$ | Specific heat capacity at constant pressure |
| $c_{v}$ | Specific heat capacity at constant volume |
| $D$ | Coefficient of diffusion |
| $d$; $d_{0}$ | Diameter |
| $E$ | Total translational energy of molecules in random motion |
| $\dot{E}$ | Average energy transferred per molecule |
| $\overline{E}$ | Average translational energy per molecule in random motion |
| $E_{m}$ | Energy transferred by molecules in the molecular range of flow |
| $E_{v}$ | Energy transferred by molecules in the viscous range of flow |

| | |
|---|---|
| $F$ | Force |
| $F_l$ | Correction factor in laminar gas flow |
| $f_v$ | Fractional number of molecules |
| G | Mass of gas incident per unit area per unit time |
| $g$ | Acceleration due to gravity |
| $H_v$ | Coefficient of heat conductivity |
| $h$ | Height |
| $J_m$ | Correction factor in molecular gas flow |
| $K_0$; $K'_0$; $K_1$; $K_2$; $K_=$ | Correction factors in molecular gas flow |
| $K_{Per}$ | Permeability coefficient |
| $K'$ | Knudsen's number |
| $k$ | Boltzmann's constant |
| $L$ | Length |
| M | Molar mass of a substance (also known as *molar weight*) |
| Ma | Mach number |
| $m$ | Mass per molecule |
| $N$ | Total number of particles in a volume |
| $N_A$ | Avogadro's constant (also known as Avogadro's "number") |
| $n$ | Number density of molecules |
| $n_M$ | Mole amount |
| $O_{ch}$ | Orifice coefficient in chocked laminar flow |
| $o$ | Perimeter |
| $p$ | Total pressure of gas |
| $\bar{p}$ | Average pressure of gas |
| $p_{a \ldots z}$; $p_{1 \ldots i}$ | Partial pressure of gas |
| $p_0$ | Pressure at the start of pumpdown |
| $p_c$ | Critical pressure |
| $p_D$ | Pressure due to outgassing of materials |
| $p_E$ | Pressure due to evaporation of materials |
| $p_G$ | Pressure due to gas load |
| $p_P$ | Pressure at the port of the pump |
| $p_{per}$ | Pressure due to permeation |
| $pr$; $p'r$ | Molecular transmission probability of a vacuum component |
| $p_s$ | Steady-state pressure |
| $p_t$ | Pressure at a certain time $t$ |
| $p_{tr}$ | Transition pressure |
| $p_u$ | Ultimate pressure of a vacuum pump |
| $p_{us}$ | Ultimate pressure of a vacuum system |
| $Q$ | Throughput |

| | |
|---|---|
| $Q_D$ | Gas flow rate due to outgassing |
| $Q_{De}$ | Gas flow rate due to outgassing of elastomers |
| $Q_{Dm}$ | Gas flow rate due to outgassing of metals |
| $Q_E$ | Gas flow rate due to evaporation of materials |
| $Q_G$ | Gas load |
| $Q_L$ | Gas flow rate due to leakage |
| $Q_{Lt}$ | Gas flow rate due to true leaks |
| $Q_{Lv}$ | Gas flow rate due to virtual leaks |
| $Q_l$ | Throughput in laminar flow of gas |
| $Q_{lch}$ | Throughput in chocked laminar flow of gas |
| $Q_{lnc}$ | Throughput in nonchocked laminar flow of gas |
| $Q_m$ | Throughput in molecular flow of gas |
| $Q_{ma}$ | Molecular throughput of gas through an aperture |
| $Q_{mp}$ | Molecular throughput of a pipe |
| $Q_r$ | Gas flow rate of residual gas |
| $Q_t$ | Throughput in turbulent flow of gas |
| $Q_{tr}$ | Throughput in transition flow of gas |
| $Q_u$ | Gas flow rate returned from the pump |
| $q_{De}$ | Specific outgassing rate of elastomers |
| $q_{Dgs}$ | Specific outgassing rate of glass and ceramics |
| $q_{Dm}$ | Specific outgassing rate of metals |
| $q_{Dpoly}$ | Specific outgassing rate of polymers |
| $q_E$ | Specific evaporation rate of a material |
| $q_e$ | Effusion rate of molecules |
| $q_{kg}$ | Mass flow rate of gas |
| $q_m$ | Molecular flow rate of gas |
| $q_{mol}$ | Molar flow rate of gas |
| $R$ | Molar constant (per "mole basis"; also known as *universal gas constant*) |
| Re | Reynold's number |
| $r$; $R'$ | Radius |
| $r_c$ | Critical pressure ratio |
| $S$ | Pumping speed—volume rate of flow |
| $S_b$ | Pumping speed of a backing pump |
| $S_m$ | Molecular pumping speed |
| $S_{ma}$ | Molecular pumping speed of an aperture |
| $S_{md}$ | Molecular pumping speed of a diaphragm |
| $S_n$ | Net (effective) pumping speed |
| $S_p$ | Pumping speed at the port of a pump |
| $S_{th}$ | Theoretical pumping speed of a pump |
| $s_f$ | Sticking coefficient |

$T$ Temperature
$T_B$ Limiting temperature to which the equation of state can be applied to high pressures
$T_c$ Critical temperature
$t$ Time
$t_m$ Time to form a monolayer
$U$ Energy
$V$ Volume
$V'$ Volume of gas at a specified temperature
$V_c$ Critical volume
$V_M$ Molar volume
$V_m$ Volume of a mixture of gases
$v$ Velocity
$v_a$ Arithmetic average velocity of molecules
$v_p$ Most probable velocity of molecules
$v_r$ Mean-square velocity of molecules
$v_s$ Velocity of sound (acoustic velocity) in gas
$W$ Electric power
$\dot{W}$ Rate of evaporation of a material
$W_t$ Mass of substance
$x_i$ Mole concentration per unit volume
$Y$ Correction factor for conductance in laminar flow
$Z$ Impedance to gas flow
$Z_T$ Impedance to gas flow of vacuum components connected in series
$\alpha$ Accommodation coefficient
$\delta_m$ Diameter (apparent) of molecule
$\varepsilon$ Collision frequency per molecule per unit time
$\eta$ Coefficient of viscosity (dynamic)
$\theta$ Slope on a log–log plot of outgassing rate versus time
$\lambda$ Mean free path of molecules in gas
$\lambda_e$ Mean free path of electrons in gas
$\lambda_i$ Mean free path of ions in gas
$\Lambda_0$ Coefficient of heat conductivity in gas
$\nu$ Ratio of specific heat capacities
$\rho$ Density of gas
$\tau$ Time constant of a vessel
$\phi$ Rate at which molecules at steady state strike unit area per unit time

# Terminology

*Accommodation coefficient* The ratio of the energy actually transferred between impinging gas molecules and the energy that theoretically would be transferred if the impinging molecules reached complete thermal equilibrium with the surface.

*Backing pressure* (also forepressure) The total pressure measured downstream from the outlet or foreline of a vacuum pump.

*Backing pump* (forepump) A vacuum pump for maintaining the backing pressure (forepressure) of another pump below its critical value.

*Bakeout* Degassing by heating materials under vacuum.

*Calibration factor* The ratio of the pressure indicated by a measuring gauge in a vacuum system to the pressure measured by a primary (or secondary) gauge in the same system.

*Celsius* The designation of the degree on the International Temperature Scale. Also used to indicate the temperature scale.

*Chocked flow* Situation encountered in the viscous flow region when the velocity of the gas through the vacuum element reaches the velocity of sound in that gas. A further reduction of the pressure downstream from the element cannot be sensed at the upstream side.

*Compression ratio* The ratio of pressures at the discharge port to the inlet port of the pump, at zero flow through the pump.

*Critical pressure* The pressure under which a gas may exist in equilibrium with its liquid phase at the critical temperature.

*Critical temperature* The temperature above which a gas cannot be liquefied by pressure alone.

*Crossover forepressure* The forepressure at which the inlet pressure of a pump becomes equal to the forepressure, so that the pump can efficiently overtake pumping.

*Crossover pressure* The pressure at which the inlet pressure of a vacuum pump becomes equal to the forepressure, or the back pressure.

*Degassing* Desorption of gas from material exposed to vacuum, as a result of supplying the material with energy such as thermal, or impact with atoms or molecules, electrons, ions, or photons.

*Desorption* The process of removing adsorbed neutral or ionized atoms and/or molecules from a material.

*Evaporation rate* The amount of material evaporated from a surface per unit area per unit time.

***Fahrenheit*** The designation of the degree and temperature scale used in English-speaking countries.

***Flow rate of gas*** The amount of moles or molecules of a certain gas species passing through a system per unit time.

***Forepressure or backpressure*** The total pressure at the outlet side of a vacuum pump, measured close to the outlet port.

***Forepressure tolerance*** (limiting forepressure, maximum forepressure, critical backpressure) The pressure at the discharge side of a vacuum pump at a stated throughput, above which the pumping action of the pump rapidly deteriorates.

***Gas ballast*** Device on oil-sealed mechanical vacuum pumps permitting the admission in the compression chamber of a certain amount of noncondensable gas (usually air) in order to maintain the saturation pressure of the vapor below the saturation value.

***Gas, ideal*** Theoretical concept that assumes that (1) the molecules are small spheres, (2) the volume of molecules is small compared to the volume actually occupied by the gas, (3) the molecules do not exert forces on each other, (4) the path traveled by molecules is linear and random, and (5) collisions between molecules are elastic.

***Gas, real*** State of a gas under given conditions of pressure and temperature when molecules exert forces on each other.

***Gas ballast*** Device on oil-sealed mechanical vacuum pumps permitting the admission to the compression chamber of a certain amount of noncondensable gas (usually air) in order to maintain the saturation pressure of the vapor below the saturation value.

***Ho coefficient*** The ratio of the measured speed of a diffusion pump to the product of its throat (or nozzle) clearance area and the maximum flow rate per unit area.

***Kelvin*** The designation of the thermodynamic temperature scale and the interval on the Practical Kelvin Temperature Scale, defined by assigning the temperature of 273.16 K to the triple point of water.

***Leak, real*** A hole or porosity in the wall of an enclosure, permitting gas to pass from one side of the wall to the other under action of a pressure or concentration differential existing across the wall.

***Leak, virtual*** The semblance of a leak from an internally trapped volume of gas (virtual volume) connected to the main vacuum system by low-conductance paths. Also an apparent leak because of the presence of contaminants that outgas very slowly within a vacuum system.

***Leak rate*** The amount of gas in pressure–volume units, at a specified temperature, passing through a leak in a given time divided by that time.

***Leak rate temperature*** The temperature of gas flowing through a leak; not to be confused with the temperature of the leak element behind the leak.

***Limiting forepressure*** (critical back pressure, tolerable forepressure) The pressure required at the outlet side of a vacuum pump. Above that pressure, the pumping action of the pump deteriorates, i.e., the inlet pressure increases.

***Mach number*** The ratio of the average flow velocity of gas across a plane and the local velocity of sound in that gas.

***Mass flow rate*** The mass of gas moving past a cross section per unit time.

***Mean free path*** The average distance a particle travels between successive collisions with other particles.

***Mercury vapor effect*** Effect that results in a pressure differential between an instrument that contains mercury at room temperature, connected to a vacuum system via a cold trap.

***Molar mass*** Mass divided by amount of substance. The SI (International System of Units) base unit is kg $mol^{-1}$ (kilograms per mole) and the practical unit is usually g $mol^{-1}$ (grams per mole).

***Mole*** The amount of substance of a system that contains as many elementary entities as there are atoms in 0.012 kg of $^{12}C$. When the mole is used, the elementary entities must be specified and may be atoms, molecules, ions, electrons, or other particles, or specified groups of such particles.

***Molecular effusion*** The molecular flow through an orifice in a wall of negligible thickness, where the largest dimension of the orifice is much smaller than the mean free path of the molecules.

***Molecular flow*** The flow of gas through a vacuum element under conditions such that the mean free path of gas molecules is considerably greater, compared to the smallest dimension of the element.

***Molecular transmission probability*** (of a vacuum element connecting two large vessels) The probability that a gas molecule entering the element from the left reaches the right-hand volume without having returned to the left-hand volume.

***Outgassing*** The spontaneous evolution of gas from materials exposed to vacuum.

***Outgassing rate (bulk)*** The quantity of gas in pressure–volume units evolved from a porous material per unit time per unit mass of material.

***Outgassing rate (surface)*** The quantity of gas in pressure–volume units evolved from the surface of a material per unit time per unit area exposed to vacuum.

***Particle*** Entity confined to a very small volume.

***Permeability coefficient*** The steady-state rate of flow of gas through a unit area and unit thickness of a solid barrier, per unit pressure differential, at a given temperature.

***Permeation*** The passage of gas through a solid barrier.

***Plenum*** A very large volume within which a gas flow will not produce a measurable pressure rise.

***Poiseuille flow*** The particular case of laminar flow through a long pipe of circular cross section.

***Pressure*** The average force that gas molecules exert on impacting on a surface. The rate of transfer of the normal component of the momentum of gas molecules with respect to a unit area, divided by that area.

***Pressure, dynamic*** The pressure indicated by a stationary gauge placed parallel to the direction of the flow.

***Pressure, static*** The pressure indicated by a gauge placed at a right angle to the direction of the flow, moving with the same velocity as the flow.

***Pressure, steady-state*** The pressure at different points within a vacuum system; remains constant with time.

***Pressure, ultimate, of a system*** The limiting pressure approached in a vacuum system after pumping time sufficient to establish that further reduction in pressure will be negligible.

***Pressure, ultimate, of a vacuum pump*** The limiting low pressure of a pump due to internal leaks of gas (e.g., gas dissolved in the pump fluid, gas from various sealing surfaces within the pump, trapped gas).

***Pumpdown, transient*** Evacuation of gas from a volume, usually at atmospheric pressure, down to the lowest pressure permitted by either the pump or by the system itself, due to the gas load.

***Pumping, steady-state*** Method to maintain a required pressure in the presence of whatever gas evolution occurs during the operation of the system.

***Pumping speed*** The ratio of the throughput of a given gas to the partial pressure of that gas at a specified point near the inlet of a pump.

***Quantity of gas*** (in pressure–volume units) The product of the pressure of gas contained in a volume, at a specified temperature, multiplied by that volume.

***Rankine*** The designation of the degree and temperature scale in English-speaking countries.

***Relative molecular mass of a substance*** (also known as *molecular mass*) The ratio of the average mass per formula unit of a substance to one half of the mass of an atom of nuclide $^{12}C$.

***Roughing out*** The removal of most gas at atmospheric pressure from a vessel (system), to the point at which a vapor pump can efficiently overtake pumping.

***Saturated vapor pressure*** State when the partial pressure equals the vapor pressure.

***Specific heat capacity*** The heat capacity per unit mass of a body, measured at constant pressure ($c_p$) or constant volume ($c_v$).

***Stationary gas*** A gas characterized by stationary distribution of velocities, i.e., where, in each elementary volume, the same number of particles can be found to have a certain velocity. The next instant, the particles in question will change their velocity as a result of collisions, but other particles will have taken this velocity, and eventually the distribution of velocities will remain the same. The gas within a closed vessel is stationary when the surface area receives and emits gas at equal rates.

***Stationary pressure*** The case when there is no net transport of energy within the mass of gas.

***Steady flow*** Flow where there is no adsorption or release of gas from or through the system or its components.

***Steady state*** The state of a dynamic vacuum system in which the mean value of parameters, e.g., gas flow, is negligible over a large period of time.

***Steady-state pumpdown*** Equilibrium obtained when the rate of gas evolved from (into) the system is equal to that at which gas is removed by the pumping system.

***Sticking coefficient*** The ratio of the number of molecules that are adsorbed on a surface for a finite period of time to the number of molecules striking the surface.

***Thermal equilibrium*** The state in which the rate at which particles enter any given distribution exactly balances the rate at which they leave.

***Thermal transpiration*** Effect that results in a pressure difference between two vessels at different temperatures, when gas flows through an element that connects the vessels, such as a porous plug or a narrow tube.

***Thin-edged aperture*** (orifice) in a flat plate The case where the wall of the flat plate is vanishingly small in the region of the aperture (orifice) so that it has no length, ideally.

***Throughput*** The amount of gas in pressure–volume units flowing per unit time across a specified cross section. The temperature of the flow must be specified. Often referred to as *pumping capacity* when measured at the pump intake.

***Time constant of a vessel*** The time required to reduce pressure at a given value in the volume of the vessel, at the existing pumping speed (volume rate of flow) of the pump that evacuates that volume. Usually expressed as "half-time" $\tau_{1/2}$ (time required to reduce pressure in a volume at half of the initial value) or $\tau_{1/10}$ (time required to reduce pressure in a volume by one decade).

***Transition flow (Knudsen flow)*** The flow of gas through a vacuum element under conditions such that the mean free path of gas molecules is of the same order of magnitude as the dimensions of the element.

***Vapor*** A gas at any temperature below its critical temperature. Such a gas cannot be liquefied by increasing pressure only. Usually referred to as *condensed gas*.

***Vapor pressure*** The pressure of a vapor in thermodynamic equilibrium with the solid or liquid at any temperature.

***Vena contracta*** The squeezing of the cross-sectional area of a gas beam as the gas enters an orifice in viscous state.

***Viscosity coefficient*** The ratio of the shearing component of stress to the velocity gradient in a gas across a plane perpendicular to the direction of the velocity gradient.

***Viscous flow*** The flow of gas through a vacuum element under conditions such that the mean free path of gas molecules is very small compared to the smallest dimension of the element. The flow may be turbulent or laminar.

***Volume rate of flow*** The volume of gas moving past a cross section, per unit time, measured at the pressure prevailing at the cross section.

***Water-vapor tolerance*** The mass flow rate for water vapor in a gas ballast pump in continuous operation and under normal ambient conditions, if the gas is water vapor.

# 1

# Ideal Gases

## 1.1 *The Ideal Gas Law*

If an ideal gas, specified by the quantities $p_1$ (pressure) and $V_1$ (volume) at temperature $T_1$, changes its state to another set of quantities $p_2, V_2$ at $T_2$, then

$$p_1V_1/T_1 = p_2V_2/T_2 \tag{1.1}$$

where $T$ is the thermodynamic temperature measured in degrees Kelvin (see Appendix, Table A.20).

## 1.2 *Boyle's Law*

For a given mass of gas, held at $T$ = const.

$$p_1V_1 = p_2V_2 = \text{const.} \tag{1.2}$$

## 1.3 *Charles's Law*

For a given mass of gas, held at $p$ = const.

$$V_1/T_1 = V_2/T_2 = \text{const.} \tag{1.3}$$

## 1.4 *Gay-Lussac's Law*

For a given mass of gas, held at $V$ = const.

$$p_1/T_1 = p_2/T_2 = \text{const.} \tag{1.4}$$

## 1.5 Mole Amount

The amount of moles $n_M$ in a given mass $W_t$ of substance having a molar mass M is

$$n_M = W_t/M \qquad [\text{mol (moles)}] \tag{1.5}$$

The molar mass M of a substance (also known as *molar weight*) is the mass divided by the *amount of substance*. The SI base unit is kg mol$^{-1}$, and the practical unit is g mol$^{-1}$ (*Compendium of Chemical Technology*, 1987, p. 260).

## 1.6 Dalton's Law

In a mixture of gases, each component exerts the pressure that it would exert if it were present alone at the same temperature in the volume occupied by the gas mixture. The total pressure $p$ of a gas mixture is the sum of partial pressures $p_1, p_2, \ldots, p_i$ of the individual components.

$$p = p_1 + p_2 + \cdots + p_i = \Sigma p_i \tag{1.6}$$

The partial pressure of each component is equal to the total pressure multiplied by its mole fraction $c_i$ in the mixture (for $c_i$, see Eq. 1.7c).

Dalton's law holds true for ideal gases. At *pressures below atmosphere, gas mixtures can be regarded as ideal gases.*

## 1.7 The Mean Molar Mass of a Mixture of Gases

If the gases in a mixture of gases have molar masses $M_1, M_2, \ldots, M_i$, then

$$\overline{M} = (W_{t1} + W_{t2} + \cdots + W_{ti})/(n_{M1} + n_{M2} + \cdots + n_{Mi}) \qquad (\text{e.g., g mol}^{-1}) \tag{1.7}$$

$$= (n_{M1}M_1 + n_{M2}M_2 + \cdots\ n_{Mi}M_i)/\Sigma n_{Mi} \tag{1.7a}$$

$$= c_1M_1 + c_2M_2 + \cdots\ c_iM_i = \Sigma c_iM_i \tag{1.7b}$$

where $c_i$ *is the mole fraction of the component* $i$ and is expressed

$$c_i = n_{Mi}/\Sigma n_{Mi} = (W_{ti}/M_i)/\Sigma(W_{ti}/M_i) = p_i/p \tag{1.7c}$$

$$\text{mol \%} = 100c_i \tag{1.7d}$$

For each component in a mixture of ideal gases, the mole percent (mol %) equals the volume percent (vol %).

**Table 1.1**
Composition of Atmosphere at Sea Level[a]

| Gas | Percent by volume | Partial pressure: Torr | Partial pressure: mbar[b] |
|---|---|---|---|
| $N_2$ | 78.08 | $5.95 \times 10^{2}$ | $7.93 \times 10^{2}$ |
| $O_2$ | 20.95 | $1.59 \times 10^{2}$ | $2.12 \times 10^{2}$ |
| Ar | 0.93 | 7.05 | 9.39 |
| $CO_2$ | $3.3 \times 10^{-2}$ | $2.5 \times 10^{-1}$ | $3.33 \times 10^{-1}$ |
| Ne | $1.8 \times 10^{-3}$ | $1.4 \times 10^{-2}$ | $1.87 \times 10^{-2}$ |
| He | $5.24 \times 10^{-4}$ | $4.0 \times 10^{-3}$ | $5.33 \times 10^{-3}$ |
| Kr | $1.1 \times 10^{-4}$ | $8.4 \times 10^{-4}$ | $1.11 \times 10^{-3}$ |
| $H_2$ | $5.0 \times 10^{-5}$ | $3.8 \times 10^{-4}$ | $5.06 \times 10^{-4}$ |
| Xe | $8.7 \times 10^{-6}$ | $6.6 \times 10^{-5}$ | $8.79 \times 10^{-5}$ |
| $H_2O$ | 1.57 | $1.29 \times 10$ | $1.72 \times 10$ |
| $CH_4$ | $2.0 \times 10^{-4}$ | $1.5 \times 10^{-3}$ | $1.99 \times 10^{-3}$ |
| $O_3$ | $7.0 \times 10^{-6}$ | $5.3 \times 10^{-5}$ | $7.06 \times 10^{-5}$ |
| $N_2O$ | $5.0 \times 10^{-5}$ | $3.8 \times 10^{-4}$ | $5.06 \times 10^{-4}$ |

[a]Adapted from Norton (1962).
[b]Values rounded down to the second digit.

Table 1.1 lists the composition of atmosphere at sea level.

The molar mass M for various gases and water-vapor values are listed in Table 1.2.

## *1.8 The Number Density n of Particles (Molecules, Atoms, etc.)*

$$n = N/V \qquad (\text{e.g., molec. m}^{-3}) \tag{1.8}$$

where $N$ is the total number of particles and $V$ the volume within which the particles are confined.

The number density $n$ can also be expressed as a function of the mole amount $n_M$. Thus

$$n = (W_t/M)(N_A/V) \qquad (\text{e.g., molec. m}^{-3}) \tag{1.8a}$$

where $N_A$ is Avogadro's number (see Eq. 1.9) and

$$n = (N_A/R)(p/T) \qquad (\text{e.g., molec. dm}^{-3}) \tag{1.8b}$$

where $R$ is the *molar gas constant*, also known as the *universal gas constant* (see Section 1.13).

**Table 1.2**

Molar Mass of Gas M, Mass of Molecules $m$, Density $\rho$ at STP, Critical Constants $T_c$ and $p_c$, and Values of the Constants $a$ and $b$ in the van der Waals Equation of State for Some Gases and Vapor[a]

| Gas | Formula | $10^3$ M ($kg\ mol^{-1}$) | $10^{26}\ m$ ($kg\ molec.^{-1}$) | $\rho$(gas) ($kg\ m^{-3}$) | $T_c$ (K) | $p_c$ (atm) | $a$ ($liters^2\ atm\ mol^{-2}$) | $b$ ($liters\ mol^{-1}$) |
|---|---|---|---|---|---|---|---|---|
| Hydrogen | $H_2$ | 2.0158 | 0.334 | 0.0899 | 33.26 | 12.8 | 0.2444 | 0.02661 |
| Nitrogen | $N_2$ | 28.0134 | 4.656 | 1.2506 | 126.16 | 33.5 | 1.390 | 0.03913 |
| Air[b] | — | 28.98 | 4.803 | 1.290 | — | — | — | — |
| Oxygen | $O_2$ | 31.9988 | 5.320 | 1.429 | 154.76 | 50.1 | 1.360 | 0.03183 |
| Helium | He | 4.0026 | 0.664 | 0.1785 | 5.26 | 2.26 | 0.03412 | 0.02370 |
| Neon | Ne | 20.183 | 3.51 | 0.9002 | 44.46 | 26.9 | 0.2107 | 0.01709 |
| Argon | Ar | 39.948 | 6.641 | 1.784 | 150.86 | 48.0 | 1.345 | 0.03219 |
| Krypton | Kr | 83.80 | 13.909 | 3.736 | 209.36 | 54.3 | 2.318 | 0.03978 |
| Xenon | Xe | 131.30 | 21.917 | 5.887 | 289.76 | 58.0 | 4.194 | 0.05105 |
| Ammonia | $NH_3$ | 17.03 | 2.870 | 0.7710 | 405.56 | 112.5 | 4.170 | 0.03707 |
| Carbon monoxide | CO | 28.01 | 4.653 | 1.250 | 133.16 | 34.5 | 1.485 | 0.03985 |
| Carbon dioxide | $CO_2$ | 44.01 | 7.360 | 1.977 | 304.16 | 72.9 | 3.592 | 0.04267 |
| Acetylene | $C_2H_2$ | 26.04 | 2.301 | 0.6181 | 308.66 | 61.6 | 4.390 | 0.05136 |
| Water | $H_2O$ | 18.0153 | 3.723 | 1.000[c] | 647.26 | 218.3 | 5.464 | 0.03049 |

[a] Values from *Handbook of Chemistry and Physics*, 67th ed. (1986–1987) for M, $\rho$ (pp. B-68–B-146), $T_c$, $p_c$ (p. F-64), $a$, $b$ (p. D-188). The values of $m$ are calculated from Eq. 1.10.
[b] Average values.
[c] Liquid.

**Table 1.3**

Values of $R$ (for 1 mol) and $k$ (for 1 molec.) in Various Systems of Units Calculated for STP from Eqs. 1.12 and 1.14

| $p$ | $V$ | $T$ | $R$ | $k$ |
|---|---|---|---|---|
| Pa (N $m^{-2}$) | $m^3$ | K | 8.314 J $mol^{-1}$ $K^{-1}$ | $1.381 \times 10^{-23}$ J $molec.^{-1}$ $K^{-1}$ |
| mbar | $dm^3$ | K | 83.14 mbar liters $mol^{-1}$ $K^{-1}$ | $1.381 \times 10^{-22}$ mbar liters $molec.^{-1}$ $K^{-1}$ |
| dyne $cm^{-2}$ | $cm^3$ | K | $8.314 \times 10^7$ ergs $mol^{-1}$ $K^{-1}$ | $1.381 \times 10^{-16}$ ergs $molec.^{-1}$ $K^{-1}$ |
| mm Hg (Torr) | $dm^3$ | K | 62.364 mm Hg liters $mol^{-1}$ $K^{-1}$ | $1.306 \times 10^{-22}$ mm Hg liters $molec.^{-1}$ $K^{-1}$ |
| atm | $cm^3$ | K | 82.058 atm $cm^3$ $mol^{-1}$ $K^{-1}$ | $1.362 \times 10^{-22}$ atm $cm^3$ $molec.^{-1}$ $K^{-1}$ |

*Note:* 1 mm Hg ≅ 1 Torr (see also footnote in Table A.5 in Appendix).

*The number density n of particles does not depend on the gas species.* The values of $n$ calculated from Eq. 1.8b are

$$n = 7.243 \times 10^{22} p/T \qquad (\text{molec. m}^{-3}) \tag{1.8c}$$

when $p$ [in pascals (Pa)], $T$ (K), $N_A = 6.022 \times 10^{23}$ molec. $mol^{-1}$, $R$ = 8.314 J $mol^{-1}$ $K^{-1}$ (see Table 1.3),

$$n = 7.243 \times 10^{21} p/T \qquad (\text{molec. dm}^{-3}) \tag{1.8d}$$

when $p$ (mbar), $T$ (K), $N_A = 6.022 \times 10^{23}$ molec. $mol^{-1}$, $R$ = 83.14 mbar liters $mol^{-1}$ $K^{-1}$ (see Table 1.3),

$$n = 9.656 \times 10^{18} p/T \qquad (\text{molec. cm}^{-3}) \tag{1.8e}$$

when $p$ (mm Hg), $T$ (K), $N_A = 6.022 \times 10^{23}$ molec. $mol^{-1}$, $R$ = 62.364 mm Hg liters $mol^{-1}$ $K^{-1}$ (see Table 1.3).

**Table 1.4**

Variation of the Number Density $n$ of Molecules for All Gases with Temperature and Pressure, Calculated from Eq. 1.8d

| $p$ (mbar) | $n$ (molec. $dm^{-3}$) $T = -195.7$°C | $T = 0$°C | $T = 23$°C |
|---|---|---|---|
| $1.0 \times 10^3$ | $9.35 \times 10^{22}$ | $2.65 \times 10^{22}$ | $2.45 \times 10^{22}$ |
| 1.0 | $9.35 \times 10^{19}$ | $2.65 \times 10^{19}$ | $2.45 \times 10^{19}$ |
| $1.0 \times 10^{-3}$ | $9.35 \times 10^{16}$ | $2.65 \times 10^{16}$ | $2.45 \times 10^{16}$ |
| $1.0 \times 10^{-6}$ | $9.35 \times 10^{13}$ | $2.65 \times 10^{13}$ | $2.45 \times 10^{13}$ |
| $1.0 \times 10^{-9}$ | $9.35 \times 10^{10}$ | $2.65 \times 10^{10}$ | $2.45 \times 10^{10}$ |
| $1.0 \times 10^{-12}$ | $9.35 \times 10^{7}$ | $2.65 \times 10^{7}$ | $2.45 \times 10^{7}$ |
| $1.0 \times 10^{-15}$ | $9.35 \times 10^{4}$ | $2.65 \times 10^{4}$ | $2.45 \times 10^{4}$ |

Table 1.5
Pressure Units

| System of units | Unit of pressure | Value |
|---|---|---|
| *Coherent* | | |
| SI (MKS) | Pa | 1 Pa = 1 N $m^{-2}$ |
| CGS | bar | 1 bar = 1.0 × $10^5$ Pa |
| MTS | pieze | 1 pz = 1.0 × $1.0^3$ Pa |
| Technical | at | 1 at = 1 kgf $cm^{-2}$ = 9.81 × $10^4$ Pa |
| British | lb $in.^{-2}$ | 1 lb $in.^{-2}$ = 6.895 × $10^3$ Pa |
| | lb $ft^{-2}$ | 1 lb $ft^{-2}$ = 4.79 × 10 Pa |
| *Noncoherent* | | |
| Physical | atm | 1 atm = 1.0 × $10^5$ Pa |
| mm of mercury[a] at 0°C | mm Hg or Torr | 1 mm Hg = 1 Torr = 1.33 × $10^2$ Pa |
| in. of mercury[a] at 32°F | in. Hg | 1 in. Hg = 3.39 × $10^3$ Pa |
| cm of water[a] at 4°C | cm $H_2O$ | 1 cm $H_2O$ = 9.80 × 10 Pa |

[a]Where the acceleration due to gravity has the standard value 9.806 m $s^{-2}$.

The values of the number density $n$ of molecules for gases at different temperatures and pressures, calculated from Eq. 1.8d, are listed in Table 1.4.

## 1.9 *Standard Conditions for Gases (STP, Standard Temperature and Pressure)*

For the calculation of measured volumes of gas, the following conditions have arbitrarily been chosen as standard:

$T$ = 0°C = 273.16 K
$p$ = 1 atm (physical atmosphere; see Table 1.5)
= 1.013250 × $10^5$ Pa = 1.013250 × $10^3$ mbar = 760 Torr

## 1.10 *Avogadro's Number*

The number of particles (molecules, atoms) in one mole of substance is given, for practical purposes, as

$$N_A = 6.022 \times 10^{23} \qquad (\text{molec. mol}^{-1}) \tag{1.9}$$

For the accurate value of $N_A$ within 7 significant digits after the decimal point, see Appendix, Table A.6.

## *1.11 The Mass per Molecule*

$$m = M/N_A \qquad (\text{e.g., g}) \tag{1.10}$$

## *1.12 The Molar Volume*

If $T = 273.16$ K and $p = 1.013250 \times 10^5$ Pa, the molar volume is

$$V_M = 22.4 \qquad (\text{liters mol}^{-1}) \tag{1.11}$$

For the accurate value of the molar volume, see Appendix, Table A.6.

## *1.13 The Equation of State*

### 1.13.1 Single Species of Gas

$$pV = n_M RT \tag{1.12}$$

$$= (W_t/M) RT \tag{1.12a}$$

where $R$ is the *molar gas constant* per "mole basis" (also known as the *universal gas constant*). The values of the molar gas constant $R$ in some system of units are listed in Table 1.3.

### 1.13.2 Mixture of Gases

If $V_m$ is the volume of the mixture and $p = p_1 + p_2 + \cdots + p_i = \Sigma p_i$ (Dalton's law; Eq. 1.6), then

$$pV_m = (\Sigma n_{Mi}) RT \tag{1.12b}$$

Equation 1.12b holds true if all the mixture components have the same pressure. Thus

$$V_m = V_1 + V_2 + \cdots + V_i = \Sigma V_i \tag{1.12c}$$

For the component $i$ in the mixture

$$p_i V_m = n_{Mi} RT \tag{1.12d}$$

Thus

$$p_i = (n_{Mi}/V_m)RT \quad (1.12e)$$

$$= x_i RT \quad (1.12f)$$

where $x_i$ is the concentration in moles per unit volume.

## 1.14 *Boltzmann's Constant*

$$k = R/N_A = 1.3806 \times 10^{-16} \quad (\text{ergs K}^{-1}) \quad (1.13)$$

$$= 1.3806 \times 10^{-23} \quad (\text{J K}^{-1}) \quad (1.13a)$$

For the conversion of *ergs* to *joules*, see Appendix, Table A.16.

For the value of $k$ accurate within 6 significant digits after the decimal point, see Appendix, Table A.6.

Boltzmann's constant $k$ is calculated per "molecular basis" in contrast with the universal gas constant $R$, which is calculated on a per "mole basis."

The values of $k$ in some system of units are listed in Table 1.3.

## 1.15 *The Gas Density*

$$\rho = mn \quad (\text{e.g., kg m}^{-3}) \quad (1.14)$$

$$= pM/RT \quad (1.14a)$$

$$= pm/kT \quad (1.14b)$$

$$= M/V_M \quad (1.14c)$$

For the expressions of $n$ and $m$, see Eqs. 1.8 and 1.10. The values of $R$ and $k$ in some system of units are listed in Table 1.3.

*The density of gas $\rho$ depends on temperature and pressure.* Therefore, these parameters must be specified when quoting the gas density.

The values of $m$ and $\rho$ (at STP) for various gases and water vapor are listed in Table 1.2.

The density of gas $\rho$ calculated from Eq. 1.14a in different systems of units is

$$\rho = 1.203 \times 10^{-4} pM/T \quad (\text{kg m}^{-3}) \quad (1.14d)$$

when $p$ (Pa or N/m$^{-2}$), $M$ (kg mol$^{-1}$), $T$ (K), $R$ (8.314 J mol$^{-1}$ K$^{-1}$; see Table 1.3),

$$\rho = 1.203 \times 10^{-2} pM/T \quad (\text{g dm}^{-3}) \quad (1.14e)$$

when $p$ (mbar), $M$ (g $mol^{-1}$), $T$ (K), $R$ (83.14 mbar $mol^{-1}$ $K^{-1}$; see Table 1.3), and

$$\rho = 1.603 \times 10^{-2} pM/T \qquad (\text{g dm}^{-3}) \qquad (1.14\text{f})$$

when $p$ (mm Hg or Torr), $M$ (g $mol^{-1}$), $T$ (K), $R$ (62.364 mm Hg $mol^{-1}$ $K^{-1}$; see Table 1.3).

## 1.16 *Gas Pressure*

The pressure $p$ exerted by gas can be expressed as

(a) The force $F$ it exerts on a unit area $A$

$$p = F/A \qquad (\text{e.g., mbar}) \qquad (1.15)$$

(b) The height $h$ of a manometric column under the acceleration $g$ due to gravity at the location of the measurement and the density $\rho$ of the liquid of the manometric column

$$p = hg\rho \qquad (\text{e.g., Torr}) \qquad (1.15\text{a})$$

(c) The number density of molecules $n$ (Eq. 1.8), Boltzmann's constant $K$ (Eq. 1.13), and the temperature at the location of the measurement $T$

$$p = nkT \qquad (\text{e.g., Pa}) \qquad (1.15\text{b})$$

(d) The number density of molecules $n$ (Eq. 1.8), the mass per molecule $m$ (Eq. 1.10), and the mean-square velocity of molecules $\bar{v}^2$ (Eq. 3.3)

$$p = (1/3)nm\bar{v}^2 \qquad (\text{e.g., } \mu\text{bar}) \qquad (1.15\text{c})$$

(e) The density of gas $\rho$ (Eq. 1.14) and the mean-square velocity of molecules $\bar{v}^2$ (Eq. 3.3)

$$p = (1/3)\rho\bar{v}^2 \qquad (\text{e.g., Torr}) \qquad (1.15\text{d})$$

The gas pressure depends only on the number density of molecules $n$ and the temperature $T$ at the location of the measurement; not on the gas species.

The unit Torr is the pressure exerted by a mercury (Hg) column 1 mm high and is defined as the 760th part of the standard atmosphere (ISO/DIS 3529/I, 1981). It differs from 1 mm Hg by 1.4 parts in $10^7$.

The micrometer (1 $\mu$m $= 10^{-3}$ mm Hg) and the millimicrometer (1 m$\mu$m $= 10^{-6}$ mm Hg) have been extensively used in the past for expressing pressure submultiples.

**Table 1.6**
Conversion Factors for Pressure in Various Systems of Units

| | Pa ($N\ m^{-2}$) | mbar | dyn $cm^{-2}$ | mm Hg (Torr) at 4°C | lb $ft^{-2}$ | Physical atmospheres (atm) | Technical atmospheres (at) | psi (lb $in.^{-2}$) | cm $H_2O$ at 4°C | in. Hg at 32°F |
|---|---|---|---|---|---|---|---|---|---|---|
| Pa | 1 | $1.0 \times 10^{-2}$ | $1.0 \times 10$ | $7.5 \times 10^{-3}$ | $2.09 \times 10^{-2}$ | $9.87 \times 10^{-6}$ | $1.02 \times 10^{-5}$ | $1.45 \times 10^{-4}$ | $1.02 \times 10^{-2}$ | $2.97 \times 10^{-4}$ |
| mbar | $1.0 \times 10^{2}$ | 1 | $1.0 \times 10^{3}$ | $7.5 \times 10^{-1}$ | 2.09 | $9.87 \times 10^{-4}$ | $1.02 \times 10^{-3}$ | $1.45 \times 10^{-2}$ | 1.02 | $2.96 \times 10^{-2}$ |
| dyn $cm^{-2}$ | $1.0 \times 10^{-1}$ | $1.0 \times 10^{-3}$ | 1 | $7.5 \times 10^{-4}$ | $2.09 \times 10^{-3}$ | $9.87 \times 10^{-7}$ | $1.02 \times 10^{-6}$ | $1.45 \times 10^{-5}$ | $1.02 \times 10^{-3}$ | $2.95 \times 10^{-5}$ |
| mm Hg (Torr) at 0°C | $1.33 \times 10^{2}$ | 1.33 | $1.33 \times 10^{3}$ | 1 | 2.78 | $1.32 \times 10^{-3}$ | $1.36 \times 10^{-3}$ | $1.93 \times 10^{-2}$ | 1.36 | $3.94 \times 10^{-2}$ |
| pound $ft^{-2}$ | $4.79 \times 10$ | $4.79 \times 10^{-1}$ | $4.79 \times 10^{2}$ | $3.59 \times 10^{-1}$ | 1 | $4.73 \times 10^{-4}$ | $4.89 \times 10^{-4}$ | $6.94 \times 10^{-3}$ | $4.89 \times 10^{-1}$ | $1.41 \times 10^{-2}$ |
| Physical atmosphere (atm) | $1.01 \times 10^{5}$ | $1.01 \times 10^{2}$ | $1.01 \times 10^{6}$ | $7.60 \times 10^{2}$ | $2.12 \times 10^{3}$ | 1 | 1.03 | $1.47 \times 10$ | $1.03 \times 10^{3}$ | $2.99 \times 10$ |
| Technical atmosphere (at) | $9.80 \times 10^{4}$ | $9.80 \times 10^{2}$ | $9.80 \times 10^{5}$ | $7.36 \times 10^{2}$ | $2.05 \times 10^{3}$ | $9.68 \times 10^{-1}$ | 1 | 1.42 | $1.0 \times 10^{3}$ | $2.89 \times 10$ |
| psi | $6.89 \times 10^{3}$ | $6.89 \times 10$ | $6.89 \times 10^{4}$ | $5.17 \times 10$ | $1.44 \times 10^{2}$ | $6.80 \times 10^{-2}$ | $7.03 \times 10^{-2}$ | 1 | $7.03 \times 10$ | 2.04 |
| cm $H_2O$ at 4°C | $9.80 \times 10$ | $9.80 \times 10^{-1}$ | $9.80 \times 10^{2}$ | $7.36 \times 10^{-1}$ | 2.05 | $9.68 \times 10^{-4}$ | $9.99 \times 10^{-4}$ | $1.42 \times 10^{-2}$ | 1 | $2.89 \times 10^{-2}$ |
| in. Hg at 32°F | $3.39 \times 10^{3}$ | $3.39 \times 10$ | $3.39 \times 10^{4}$ | $2.54 \times 10$ | $7.07 \times 10$ | $3.34 \times 10^{-2}$ | $3.45 \times 10^{-2}$ | $4.91 \times 10^{-1}$ | $3.45 \times 10$ | 1 |

**Table 1.7**
Classification of Vacuum Ranges[a]

| | Pressure units | | |
|---|---|---|---|
| | Pa | mbar | Torr |
| Low | $3.3 \times 10^3 < p < 1.0 \times 10^5$ | $3.3 \times 10 < p < 1.0 \times 10^3$ | $25.0 < p < \sim 750$ |
| Medium | $1.0 \times 10^{-1} < p \leqslant 3.3 \times 10^3$ | $1.0 \times 10^{-3} < p \leqslant 3.3 \times 10$ | $7.5 \times 10^{-4} < p \leqslant 25.0$ |
| High | $1.0 \times 10^{-4} < p \leqslant 1.0 \times 10^{-1}$ | $1.0 \times 10^{-6} < p \leqslant 1.0 \times 10^{-3}$ | $7.5 \times 10^{-7} < p \leqslant 7.5 \times 10^{-4}$ |
| Very high | $1.0 \times 10^{-7} < p \leqslant 1.0 \times 10^{-4}$ | $1.0 \times 10^{-9} < p \leqslant 1.0 \times 10^{-6}$ | $7.5 \times 10^{-10} < p \leqslant 7.5 \times 10^{-7}$ |
| Ultrahigh | $1.0 \times 10^{-10} < p \leqslant 1.0 \times 10^{-7}$ | $1.0 \times 10^{-12} < p \leqslant 1.0 \times 10^{-9}$ | $7.5 \times 10^{-13} < p \leqslant 7.5 \times 10^{-10}$ |
| Extreme ultrahigh | $p \leqslant 1.0 \times 10^{-10}$ | $p \leqslant 1.0 \times 10^{-12}$ | $p \leqslant 7.5 \times 10^{-13}$ |

[a]Adapted from AVS *Dictionary for Vacuum Science and Technology* (1980).

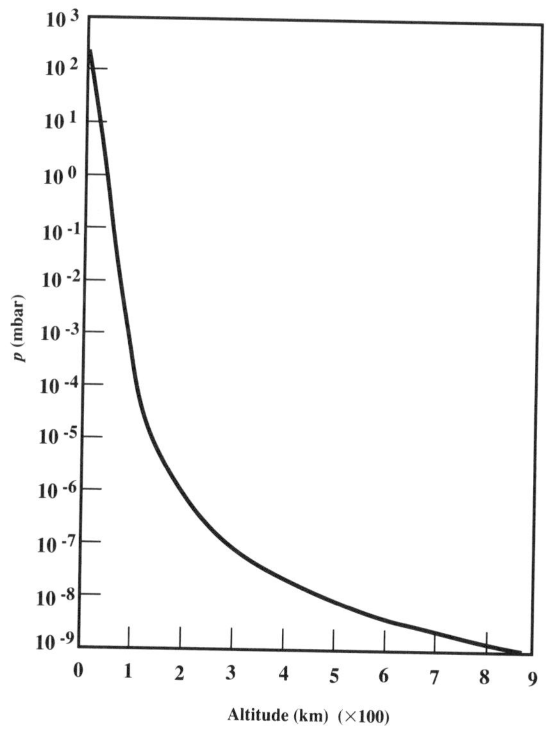


**Figure 1.1** Variation in atmospheric pressure with altitude.

The pressure units in coherent and noncoherent systems as well as the conversion factors for pressure in various systems of units are listed in Tables 1.5 and 1.6.

For conversion factors for pressure, accurate within 5 significant digits after the decimal point, as well as for other units not specified in Table 1.6, see Moss (1987).

Vacuum is the gaseous environment at pressures below atmosphere, and is conventionally classified in the ranges listed in Table 1.7.

The variation of the atmospheric pressure with altitude is illustrated in Fig. 1.1.

## *Exercises*

**Exercise 1.1**

A calibration system contains gas at a temperature $T = 23°C$ and a pressure $p = 800$ mbar. Determine the percent increase of the pressure in the system after the temperature increases to 26°C. Assume that at both temperatures the gas is in thermal equilibrium with the walls of the system.

**Solution 1.1**

Since the system operates at constant volume, use Eq. 1.4, $p_2 = p_1(T_2/T_1)$. Knowing that $T_K = 273 + T_{°C}$, substitute

$T_1 = 273 + 23 = 296$ K
$T_2 = 273 + 26 = 299$ K
$p = 800$ mbar

There results

$$p_2 = 800 \text{ mbar} \times (299 \text{ K}/296 \text{ K}) = 808.108 \text{ mbar}$$

The increase is

$$100 \times (808.108 \text{ mbar}/800 \text{ mbar}) = 1.01\%$$

**Exercise 1.2**

Gas at a pressure of 200 kPa and a temperature of 32°F occupies a volume of 353 $ft^3$. Determine the volume occupied by gas at a temperature of 212°F and a pressure of 500 kPa.s

**Solution 1.2**

Knowing that $T_K = (T_{°F} + 460)(5/9)$ (see Appendix, Table A.20) and using Eq. 1.1

$$V_2 = (p_1V_1/T_1)(T_2/p_2)$$

one substitutes

$p_1 = 200$ kPa
$p_2 = 500$ kPa
$V_1 = 353$ ft$^3$
$T_1 = (32 + 460)(5/9) = 273$ K
$T_2 = (212 + 460)(5/9) = 373$ K
$V_2 = (200 \text{ kPa} \times 353 \text{ ft}^3/273 \text{ K})(373 \text{ K}/500 \text{ kPa}) = 193 \text{ ft}^3$

**Exercise 1.3**

The manometer on a gas cylinder, stored for a long time at a temperature of 15°C, reads a pressure of 1176 psi (lb in.$^{-2}$). What would be the absolute pressure of gas in the cylinder, if the cylinder were shifted to a place where its temperature reaches 35°C. The atmospheric pressure at the moment of reading the manometer is 14 psi. Assume that the cylinder is tight so that the mass of gas did not change and that the gas has reached thermal equilibrium with the walls.

**Solution 1.3**

The pressure reading on usual manometers used on gas cylinders is the *gauge pressure*, i.e., *the actual gas pressure in the cylinder + the atmospheric pressure.* Since the gas cylinder is tight, $V_{\text{cylinder}}$ = const., and Eq. 1.4 applies. Thus

$$p_2 = p_1(T_2/T_1)$$

Substitute

$p_{1,\text{ gauge}} = 1176$ psi
$p_{1,\text{ actual}} = 1176 \text{ psi} - 14 \text{ psi} = 1162$ psi
$T_1 = 273 + 15 = 288$ K
$T_2 = 273 + 35 = 308$ K

Thus

$$p_{2,\text{ gauge}} = 1176 \text{ psi} \times (308 \text{ K}/288 \text{ K}) = 1257 \text{ psi}$$

$$p_{2,\text{ actual}} = 1257 \text{ psi} - 14 \text{ psi} = 1243 \text{ psi}$$

**Exercise 1.4**

A mass of gas at an absolute pressure of 120 atm occupies a volume of 43 liters. Find the volume of gas at 1 atm, assuming that the temperature remains constant.

**Solution 1.4**

Use Eq. 1.2:

$$V_2 = (p_2/p_1)V_1$$

and substitute

$p_1 = 120$ atm
$p_2 = 1$ atm
$V_1 = 43$ liters

Thus

$$V_2 = (120 \text{ atm}/1 \text{ atm}) \times 43 \text{ liters} = 5160 \text{ liters}$$

**Exercise 1.5**

What are the partial pressures, in millibars, of oxygen and nitrogen in a dry mixture of 21% $O_2$ and 79% $N_2$ at 1 atm?

**Solution 1.5**

Use Eq. 1.7c written as

$$p_i = pc_i$$

and substitute

$c_{O_2} = 21\%\ O_2/(21\%\ O_2 + 79\%\ N_2)$
$c_{N_2} = 79\%\ N_2/(21\%\ O_2 + 79\%\ N_2)$
$p = 1 \text{ atm} = 1.01 \times 10^3$ mbar (see Table 1.6)

Then

$p_{O_2} = 1.01 \times 10^3 \text{ mbar} \times 21\%\ O_2/(21\%\ O_2 + 79\%\ N_2) = 212.1$ mbar
$p_{N_2} = 1.01 \times 10^3 \text{ mbar} \times 79\%\ N_2/(21\%\ O_2 + 79\%\ N_2) = 797.9$ mbar

**Exercise 1.6**

The main components in percent by volume of dry atmosphere at sea level are $N_2 = 78.08$, $O_2 = 20.95$, $Ar = 0.93$, $CO_2 = 3.3 \times 10^{-2}$, $Ne = 1.8 \times 10^{-3}$ (see Table 1.1). Calculate (a) the mean molecular mass $\overline{M}$ and (b) the density of air having this composition.

**Solution 1.6**

(a) Since for each component in a mixture of ideal gases, the mol% = vol% (see Section 1.7), Eq. 1.7b is used: $\overline{M} = \Sigma c_i M_i$. On substituting the values of the molecular masses (rounded-down values; see Table 1.2)

$M_{N_2} = 28.0$ g mol$^{-1}$
$M_{O_2} = 32.0$ g mol$^{-1}$
$M_{Ar} = 40.0$ g mol$^{-1}$
$M_{CO_2} = 44.0$ g mol$^{-1}$
$M_{Ne} = 20.2$ g mol$^{-1}$

the value of $\overline{M}$ results as

$$\overline{M} = (1/100)(78.08 \times 28.0 + 20.95 \times 32 + 0.93 \times 40.0 + 3.3 \times 10^{-2} \times 44.0 + 1.8 \times 10^{-3} \times 20.2)\ \text{g mol}^{-1}$$

$$= 28.95\ \text{g mol}^{-1}$$

(b) Substitute in Eq. 1.14c

$$\rho = M/V_M$$
$$M = \overline{M} = 28.95\ \text{g mol}^{-1}$$
$$V_M = 22.4\ \text{liters mol}^{-1}\ \text{(see Eq. 1.11)}$$

Thus

$$\rho = 28.95\ \text{g mol}^{-1}/22.4\ \text{liters mol}^{-1} = 1.29\ \text{g liter}^{-1}$$

**Exercise 1.7**

What are the values of (a) the molar gas constant (universal gas constant) $R$ and (b) Boltzmann's constant $k$ in the CGS (centimeter–gram–second) system of units for 1 mol of substance?

**Solution 1.7**

The values of both $R$ and $k$ in the CGS system of units are calculated at STP (see Section 1.9), where

$$T = 273.16\ \text{K}$$
$$p = 1.013250 \times 10^5\ \text{Pa} = 1.013250 \times 10^6\ \text{dyn cm}^{-2}\ \text{(see Table 1.7)}$$

(a) Equation 1.12 is used:

$$pV = n_M RT$$

where $n_M = 1$ mol, $V = V_M = 22414.10\ \text{cm}^3\ \text{mol}^{-1}$ (see Section 1.12 and Appendix, Table A.6). Thus

$$R = (1.013250 \times 10^6\ \text{dyn cm}^{-2} \times 22414.10\ \text{cm}^3\ \text{mol}^{-1})/(273.16\ \text{K})$$

$$= 8.3142066 \times 10^7\ \text{dyn cm mol}^{-1}\ \text{K}^{-1} = 8.3142066 \times 10^7\ \text{erg mol}^{-1}\ \text{K}^{-1}$$

(b) Substitute in Eq. 1.13

$$k = R/N_A$$

the value of $R$ calculated at (a) and $N_A = 6.0221367 \times 10^{23}$ molec. $\text{mol}^{-1}$ (see Appendix, Table A.6).

$$k = 8.3142066 \times 10^7\ \text{erg mol}^{-1}\ \text{K}^{-1}/6.0221367 \times 10^{23}\ \text{molec. mol}^{-1}$$

$$= 1.380658 \times 10^{-16}\ \text{erg K}^{-1}\ \text{molec.}^{-1}$$

**Exercise 1.8**

What is the volume of 1 mol at STP *for any gas*?

**Solution 1.8**

Substitute in Eq. 1.12 written as

$$V = n_M RT/p$$

the following values

$$\begin{aligned} n_M &= 1 \text{ mol} \\ R &= 8.314 \text{ J mol}^{-1} \text{ K}^{-1} \text{ (see Table 1.3)} \\ p &= 1.013250 \times 10^5 \text{ Pa (see Section 1.9)} \\ T &= 273.16 \text{ K} \end{aligned}$$

Then

$$\begin{aligned} V &= 1 \text{ mol} \times 8.314 \times 10^7 \text{ J mol}^{-1} \text{ K}^{-1} \times 273.16 \text{ K}/1.013250 \times 10^5 \text{ Pa} \\ &= 1 \text{ mol} \times 8.314 \times 10^7 \text{ N m mol}^{-1} \text{ K}^{-1} \\ &\quad \times 273.16 \text{ K}/1.013250 \times 10^5 \text{ N m}^{-2} \\ &= 22.41 \text{ liters} \end{aligned}$$

**Exercise 1.9**

A vacuumtight chamber of 0.02 $dm^3$, evacuated well below the atmospheric pressure, contains 2 mg of water vapor. What pressure would result in the chamber when backing it to 400°C? The temperature of the chamber before backing was 23°C. Assume that the water vapor is an ideal gas.

**Solution 1.9**

The pressure rise, due to the water vapor, after backing is calculated from Eq. 1.12a:

$$p = (W_t/M)(RT/V)$$

Substitute

$$\begin{aligned} W_t &= 2.0 \times 10^{-6} \text{ kg water vapor} \\ M &= 18.02 \times 10^{-3} \text{ kg mol}^{-1} \text{ (see Table 1.2)} \\ R &= 8.314 \text{ J mol}^{-1} \text{ K}^{-1} \text{ (see Table 1.3)} \\ T &= 400°\text{C} - 23°\text{C} = 377°\text{C} = 273 + 377 = 650 \text{ K} \\ V &= 2.0 \times 10^{-5} \text{ m}^3 \end{aligned}$$

The pressure rise is

$$\begin{aligned} p &= (2.0 \times 10^{-6} \text{ kg}/18.02 \times 10^{-3} \text{ kg mol}^{-1}) \\ &\quad \times (8.314 \text{ J mol}^{-1} \text{ K}^{-1} \times 650 \text{ K}/2.0 \times 10^{-5} \text{ m}^3) \\ &= 2.9 \times 10^4 \text{ N mm}^{-3} \\ &= 2.9 \times 10^4 \text{ N m}^{-2} = 2.9 \times 10^4 \text{ Pa} \end{aligned}$$

**Exercise 1.10**

Determine the volume occupied by 2 g of helium at 23°C and 100 Torr.

**Solution 1.10**

Use Eq. 1.12a written as

$$V = (W_t/M)(RT/p)$$

and substitute

$W_t = 2$ g
$M_{He} = 4.0$ g mol$^{-1}$ (see Table 1.2)
$R = 62.364$ Torr liters mol$^{-1}$ K$^{-1}$ (see Table 1.3)
$T = 273 + 23 = 296$ K
$p = 100$ Torr

There results

$$V = (2 \text{ g}/4.0 \text{ g mol}^{-1})(62.364 \text{ Torr liters mol}^{-1} \text{ K}^{-1} \times 296 \text{ K}/100 \text{ Torr})$$
$$= 92.3 \text{ liters}$$

**Exercise 1.11**

A gas cylinder having an empty weight of 81 kg and an internal volume of 50 liters is filled with nitrogen at a gauge pressure of 170 atm. At the moment of the filling the ambient temperature was 20°C and the air pressure was 14 psi. How many kilograms of nitrogen can carry the cylinder?

**Solution 1.11**

Substitute in Eq. 1.12a

$$W_t = MpV/RT$$

the following data

$M_{N_2} = 28.0 \times 10^{-3}$ kg mol$^{-1}$ (see Table 1.2)
$R = 82.058$ atm cm$^3$ mol$^{-1}$ K$^{-1}$ (see Table 1.3)
$T = 273 + 20 = 293$ K
$p_{absolute} = p_{gauge} - p_{atm}$ (see Exercise 1.3)
$= 170$ atm $- (14$ psi $\times 6.8 \times 10^{-2}$ atm/psi)
$= 169$ atm (for the conversion of psi to atm, see Table 1.6)

$$V = 50 \text{ liters} = 50 \text{ dm}^3$$

There results

$$W_t = (28.0 \times 10^{-3} \text{ kg mol}^{-1} \times 169 \text{ atm} \times 50 \text{ dm}^3)/$$
$$(82.058 \times 10^{-3} \text{ atm dm}^3 \text{ mol}^{-1} \text{ K}^{-1} \times 293 \text{ K})$$
$$= 9.84 \text{ kg}$$

(*Note*: Usually the weight of an empty cylinder is about 90% of the carriage weight.)

**Exercise 1.12**

A gas cylinder of 50-liter capacity contains argon at a temperature of 23°C and a gauge pressure of 300 psi. As argon is withdrawn from the cylinder, the gauge pressure drops to 80 psi and the gas temperature decreases to 10°C. Determine

(a) The amount of moles and the mass of argon initially available in the cylinder.
(b) The mass, in kilograms, of argon withdrawn.
(c) The volume of the gas withdrawn from the cylinder at a pressure of 1 atm and a temperature of 23°C.

**Solution 1.12**

(a) The initial amount of argon is calculated from Eq. 1.12a:

$$W_t/M_{Ar} = pV/RT$$

Substitute

$W_{t1}$ = the initial mass of argon in the cylinder
$p_1$ = 21.42 atm, the initial pressure of argon calculated as

$$p_1 = p_{1,\text{gauge}} + p_{\text{atm}} = (300 + 15)\ \text{psi} \times 6.8 \times 10^{-2}\ \text{atm psi}^{-2}$$

(for converting psi to atm, see Table 1.6)
$V_{cyl} = 50\ \text{liters} = 50 \times 10^3\ \text{cm}^3$
$R = 82.058\ \text{atm cm}^3\ \text{mol}^{-1}\ \text{K}^{-1}$ (see Table 1.3)
$T_1 = 273 + 23 = 296$ K, the initial temperature of argon
$M_{Ar} = 39.9\ \text{g mol}^{-1}$ (see Table 1.2)

Therefore

$$\begin{aligned} W_{t1}/M_{Ar} &= (21.42\ \text{atm} \times 50 \times 10^3\ \text{cm}^3)/ \\ &\quad (82.058\ \text{atm cm}^3\ \text{mol}^{-1}\ \text{K}^{-1} \times 296\ \text{K}) \\ &= 44.09\ \text{mol} \end{aligned}$$

and the mass of argon initially available in the cylinder

$$\begin{aligned} W_{t1} &= (W_{t1}/M_{ar})M_{Ar} = 44.09\ \text{mol} \times 39.9\ \text{g mol}^{-1} \\ &= 17.59\ \text{kg} \end{aligned}$$

(b) To calculate the mass of argon withdrawn, first determine the mass of argon remaining in the cylinder. Use Eq. 1.12a as written in (a) and substitute

$W_{t_2}$ = the mass of argon that remains in the cylinder
$p_2$ = 6.46 atm the pressure of argon remaining in the cylinder calculated from
$p_2 = p_{2,\text{gauge}} + p_{\text{atm}} = (80 + 15)\ \text{psi} \times 6.8 \times 10^{-2}\ \text{atm psi}^{-2}$
$v_{\text{cyl}}$ = 50 liters
$R$ = 82.058 atm cm$^3$ mol$^{-1}$ K$^{-1}$ (see Table 1.3)
$T_2 = 273 + 10 = 283$ K

Therefore

$$W_{t2}/M_{\text{Ar}} = (6.46\ \text{atm} \times 50 \times 10^3\ \text{cm}^3)/$$
$$(82.058\ \text{atm cm}^3\ \text{mol}^{-1}\ \text{K}^{-1} \times 283\ \text{K}) = 13.9\ \text{mol}$$

and the mass of argon remaining in the cylinder

$$W_{t2} = (W_{t2}/M_{\text{Ar}})M_{\text{Ar}} = 13.9\ \text{mol} \times 39.9\ \text{kg mol}^{-1} = 0.555\ \text{kg}$$

The mass of the gas withdrawn is

$$W_{t1} - W_{t2} = 9.84\ \text{kg} - 0.555\ \text{kg} = 9.285\ \text{kg}$$

(c) Using Eq. 1.12a, determine the volume occupied by the gas withdrawn:

$$V_{\text{Ar}} = ((W_{t1}/M_{\text{Ar}}) - (W_{t2}/M_{\text{Ar}}))RT_1/p$$

Substitute

$W_{t1}/M_{\text{Ar}}$ = 44.09 mol
$W_{t2}/M_{\text{Ar}}$ = 13.9 mol
$R$ = 82.058 × 10$^3$ atm cm$^3$ mol$^{-1}$ K$^{-1}$ (see Table 1.3)
$T_1$ = 273 + 23 = 296 K
$p$ = 1 atm

Thus

$$V_2 = (44.09\ \text{mol} - 13.9\ \text{mol}) \times 82.058 \times 10^{-3}\ \text{atm liters mol}^{-1}\ \text{K}^{-1}$$
$$\times 296\ \text{K}/1\ \text{atm} = 733\ \text{liters}$$

**Exercise 1.13**

What is (a) the mean molecular mass of a mixture of 50 g of nitrogen at 143.75 mbar, 108 g of oxygen at 271.5 mbar, 42 g of argon at 84.5 mbar, and (b) the mole fraction $c$ of each component.

**Solution 1.13**

(a) Use Eq. 1.5

$$n_M = W_t/M$$

and Eq. 1.7

$$\overline{M} = (W_{t1} + W_{t2} + \cdots + W_{ti})/(n_{M1} + n_{M2} + \cdots + n_{Mi})$$

to determine the mean molecular mass $\overline{M}$. Substitute the following data:

$$W_{t(N_2)} = 50 \text{ g} \qquad M_{N_2} = 28 \text{ g mol}^{-1} \text{ (see Table 1.2)}$$

$$W_{t(O_2)} = 108 \text{ g} \qquad M_{O_2} = 32 \text{ g mol}^{-1} \text{ (see Table 1.2)}$$

$$W_{t(Ar)} = 42 \text{ g} \qquad M_{Ar} = 40 \text{ g mol}^{-1} \text{ (see Table 1.2)}$$

There results

$$n_{M(N2)} = 50 \text{ g}/28 \text{ g mol}^{-1} = 1.786 \text{ mol}$$
$$n_{M(O2)} = 108 \text{ g}/32 \text{ g mol}^{-1} = 3.375 \text{ mol}$$
$$n_{M(Ar)} = 42 \text{ g}/40 \text{ g mol}^{-1} = 1.05 \text{ mol}$$
$$\overline{M} = (50 \text{ g} + 108 \text{ g} + 42 \text{ g})/(1.786 \text{ mol} + 3.375 \text{ mol} + 1.05 \text{ mol}) = 32.2 \text{ g mol}^{-1}$$

(b) Substitute in Eq. 1.7c

$$c_i = p_i/p$$

the following data

$$p_{N_2} = 143.75 \text{ mbar}$$
$$p_{O_2} = 271.5 \text{ mbar}$$
$$p_{Ar} = 84.5 \text{ mbar}$$
$$p = 143.75 \text{ mbar} + 271.5 \text{ mbar} + 84.5 \text{ mbar} = 499.75 \text{ mbar}$$

Thus, the mole fraction $c$ of each component of the mixture results as

$$c_{N_2} = 143.75 \text{ mbar}/499.75 \text{ mbar} = 0.288$$
$$c_{O_2} = 217.5 \text{ mbar}/499.75 \text{ mbar} = 0.543$$
$$c_{Ar} = 84.5 \text{ mbar}/499.75 \text{ mbar} = 0.169$$

**Exercise 1.14**

Determine the volume and the density of 1 g of carbon dioxide at 23°C and 0.1 mbar. Compare the value found to that listed in Table 1.2.

**Solution 1.14**

Use Eq. 1.12a

$$V = (W_t/M)RT/p$$

and Eq. 1.14c

$$\rho = M/V_M$$

Substitute

$W_{t(CO_2)} = 1$ g
$M_{(CO_2)} = 44.01$ g mol$^{-1}$ (see Table 1.2)
$R = 83.14$ mbar liters mol$^{-1}$ K$^{-1}$ (see Table 1.3)
$T = 273 + 23 = 296$ K
$p = 0.1$ mbar
$V_M = 22.4141$ liters mol$^{-1}$ (see Appendix, Table A.6)

Thus

$$V = (1 \text{ g}/44.01 \text{ g mol}^{-1}) \times 83.14 \text{ mbar liters mol}^{-1} \text{ K}^{-1}$$
$$\times 296 \text{ K}/0.1 \text{ mbar}$$
$$= 5592 \text{ liters} = 5.592 \text{ m}^3$$
$$\rho = 44.01 \text{ g mol}^{-1}/22.4141 \text{ liters mol}^{-1} = 1.963 \times 10^{-3} \text{ kg m}^3$$

The density in Table 1.2 was calculated from Eq. 1.14b

$$\rho = pm/kT$$

for STP conditions. On substituting

$p = 1.013250 \times 10^3$ mbar (see Section 1.9)
$m = 7.360 \times 10^{-26}$ kg molec$^{-1}$ (see Table 1.2)
$k = 1.381 \times 10^{22}$ mbar liters molec.$^{-1}$ K$^{-1}$ (see Table 1.3)
$T = 273.16$ K

There results

$$\rho = (1.013250 \times 10^3 \text{ mbar} \times 7.360 \times 10^{-26} \text{ kg molec.}^{-1})/$$
$$(1.381 \times 10^{-22} \text{ mbar molec.}^{-1} \text{ K}^{-1} \times 273.16 \text{ K})$$
$$\rho = 1.977 \times 10^{-3} \text{ kg m}^{-3}$$

**Exercise 1.15**

Determine the density of xenon at STP given that the density of air at STP is 1.290 kg m$^{-3}$.

**Solution 1.15**

Equation 1.14c written for xenon and air is

$$\rho_{Xe} = M_{Xe}/V_M$$
$$\rho_{air} = M_{air}/V_M$$

On dividing $\rho_{Xe}/\rho_{air}$ there results

$$\rho_{Xe} = \rho_{air}(M_{Xe}/M_{air})$$

Substitute in this equation

$\rho_{air} = 1.290 \times 10^{-3}$ kg m$^{-3}$ (see Table 1.2)
$M_{air} = 28.98 \times 10^{-3}$ kg mol$^{-1}$ (see Table 1.2)
$M_{Xe} = 131.30 \times 10^{-3}$ kg mol$^{-1}$ (see Table 1.2)

Thus

$$\rho_{Xe} = 1.290 \text{ kg m}^{-3} \times 141.3 \times 10^{-3} \text{ kg mol}^{-1}/28.98 \times 10^{-3} \text{ kg mol}^{-1}$$
$$= 5.84 \times 10^{-3} \text{ kg m}^{-3}$$

(*Note*: The method of determining the density of a gas by relating it to the density of air is accurate within 4%.)

**Exercise 1.16**

Convert pressure measured in millimeters of mercury (mm Hg) (Torr) to pascals (N m$^{-2}$).

**Solution 1.16**

Use Eq. 1.15a:

$$p = hg\rho$$

Substitute

$h$ = 1 mm (the height of the barometric mercury column)
$g$ = 9.80665 m s$^{-2}$ (internationally accepted value for the acceleration due to gravity; see Appendix, Table A.7)
$\rho$ = 13.59508 kg dm$^{-3}$ (mercury density, see Appendix, Table A.7)

Calculate $p$ in the CGS system:

$$p = 0.1 \text{ cm} \times 980.665 \text{ cm s}^{-2} \times 13.59508 \text{ g cm}^{-3} = 1333.2219 \text{ g cm}^{-1} \text{ s}^{-2}$$

Since 1 Pa = 1 N m$^{-2}$ = 1 kg m s$^{-2}$ $\times$ m$^{-2}$ = 1 kg m$^{-1}$ s$^{-2}$

$$p = 1.33322 \times 10^{2} \text{ Pa}$$

**Exercise 1.17**

What is the mass of air at atmospheric pressure and 23°C in a room of 25 m$^3$?

**Solution 1.17**

The total number of molecules $N$ is calculated from Eq. 1.8 written as

$$N = nV$$

and Eq. 1.18c

$$n = 7.243 \times 10^{22}\ p_{\mathrm{Pa}}/T_{\mathrm{K}}$$

Then the mass of air $W_t$ in the room is determined as

$$W_t = Nm = nVm = \left(7.234 \times 10^{22}\ p_{\mathrm{Pa}}/T_{\mathrm{K}}\right) \times mV$$

Substitute

$V = 25\ \mathrm{m}^3$
$p = 1\ \mathrm{atm} = 1.01 \times 10^5\ \mathrm{Pa}$ (see Table 1.6)
$T = 273 + 23 = 296\ \mathrm{K}$
$m = 4.8 \times^{-26}\ \mathrm{kg\ molec.}^{-1}$ (see Table 1.2)

Thus

$$W_t = \left(7.243 \times 10^{22} \times 1.01 \times 10^5/296\right) \mathrm{molec.\ m}^{-3} \times 25\ \mathrm{m}^3$$
$$\times 4.8 \times 10^{-6}\ \mathrm{kg\ molec.} = 29.7\ \mathrm{kg}$$

**Exercise 1.18**

Determine the number of molecules of gas composing the atmospheric air in 1 cm$^3$ at STP.

**Solution 1.18**

The number of gas molecules in 1 cm$^3$ is computed from Eq. 1.8b, which gives the number density of molecules per unit volume

$$n = (N_A/R)(p/T)$$

Substitute

$N_A = 6.022 \times 10^{23}\ \mathrm{molec.\ mol}^{-1}$ (see Appendix Table A.6)
$R = 82.058\ \mathrm{atm\ cm}^3\ \mathrm{mol}^{-1}\ \mathrm{K}^{-1}$ (see Table 1.3)
$p = 1\ \mathrm{atm}$
$T = 273.16\ \mathrm{K}$

Then

$$n = \left(6.022 \times 10^{23}\ \mathrm{molec.\ mol}^{-1}/82.058\ \mathrm{atm\ cm}^3\ \mathrm{mol}^{-1}\ \mathrm{K}^{-1}\right)$$
$$\times (1\ \mathrm{atm}/273.16) = 2.686 \times 10^{19}\ \mathrm{molec.\ cm}^{-3}$$

[*Note*: This value is known as *Loschmidt's number* (*Loschmidt's constant*).]

**Exercise 1.19**

In a barometer made of a tube of glass of 100-cm length and 13-mm bore, the mercury column stands at a height of 75 cm when the ambient temperature is 20°C. As a result of a crack in the glass tube above the mercury, air leaks into the empty space of the barometer and depresses the column to a height of 68 cm. Determine the mass of air leaked into the barometer on disregarding the pressure due to mercury vapor above the mercury meniscus. Assume that the ambient temperature remains constant.

**Solution 1.19**

Referring to Fig. 1A,

(a) Calculate the barometric pressure due to the air leak, from Eq. 1.15a, written as

$$p = (h_1 - h_2)g$$

Substitute

$h_1$ = 0.75 m (the initial height of the mercury column)
$h_2$ = 0.68 m (the height of the mercury column after the leak)
$g$ = 9.81 m s$^{-2}$ (see Appendix, Table A.7)
$\rho$ = 13.6 × 10$^3$ kg m$^{-3}$ (see Table A.7)

Thus

$$p = (0.75 - 0.68)\ \text{m} \times 9.81\ \text{m s}^{-2} \times 13.6 \times 10^3\ \text{kg m}^{-3}$$
$$= 9.339 \times 10^3\ \text{kg m s}^{-2} = 9.339 \times 10^3\ \text{Pa}$$

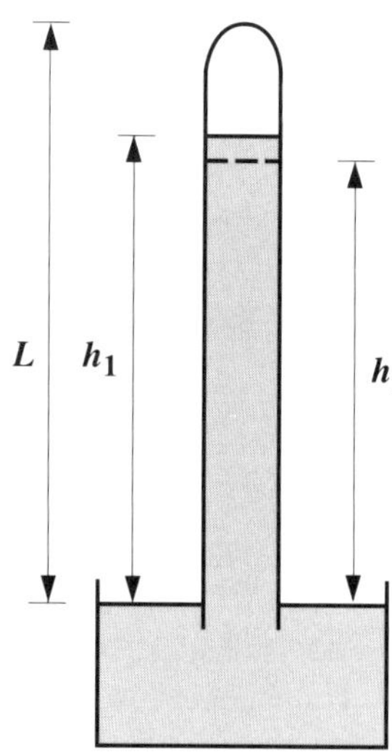


**Figure 1A**

(b) Determine the volume occupied by air in the barometer, assuming that the sealing of the glass tube above the mercury column is flat (see Berman, 1985, p. 77).

The volume of the barometric tube is

$$V = (\pi d^2/4)(Lh_2)$$

Substitute

$d = 0.013$ mm (the bore of the glass tube)
$L = 1.0$ m (the length of the glass tube)

Thus

$$(V = 3.14 \times 0.013 \text{ m}^2/4)(1.0 \text{ m} - 0.68 \text{ m}) = 4.25 \times 10^{-5} \text{ m}^3$$

(c) Calculate the number density $n_V$ of molecules within the volume $V$ occupied by air:

$$n_V = nV$$

where $n$ is determined from Eq. 1.8c:

$$n = 7.243 \times 10^{22} p_{\text{Pa}}/T_{\text{K}}$$

Thus

$$n = 7.243 \times 10^{22} p_{\text{Pa}}/T_{\text{K}} \text{ (molec. m}^{-3})V(\text{m}^3)$$

Substitute

$p = 9.339 \times 10^3$ Pa [see (a)]
$T = 273 + 20 = 293$ K
$V = 4.25 \times 10^{-5}$ m$^3$ [see (b)]

There results

$$n_V = 7.243 \times 10^{22} \times (9.339 \times 10^3/293) \text{ molec. m}^{-3}$$
$$\times 4.25 \times 10^{-5} \text{ m}^3 = 9.81 \times 10^{19} \text{ molec.}$$

(d) The mass of air $W_{\text{t, air}}$ leaked in the barometer is

$$W_{\text{t, air}} = n_V m_{\text{air}}$$

Substitute

$n_V = 9.81 \times 10^{19}$ molec. [see (c)]
$m_{\text{air}} = 4.8 \times 10^{-26}$ kg molec.$^{-1}$ (see Table 1.2)

Thus

$$W_{t,\,air} = 9.81 \times 10^{19}\ \text{molec.} \times 4.8 \times 10^{-26}\ \text{kg molec.}$$

$$= 4.7 \times 10^{-6}\ \text{kg} = 4.7\ \text{mg}$$

**Exercise 1.20**

Estimate the order of magnitude of the distance between molecules at 1013 mbar and 23°C.

**Solution 1.20**

As a general rule, the number of objects $n_{obj}$ of any kind within a given volume is inversely proportional to the cube of the average distance $L$ that separates the objects. Thus

$$n_{obj} = L^{-3} \qquad \left(\text{e.g., } n_{obj}\ \text{cm}^{-3}\right)$$

*This relationship holds true if, and only if, the objects are equidistant and do not change their position with time.*

Assuming that the molecules within a certain volume are steady and equidistant (actually not true; see Chapter 3), the number density $n$ of molecules is calculated from Eq. 1.8d as

$$n = 7.243 \times 10^{21} p_{mbar}/T_K \qquad (\text{molec. dm}^{-3})$$

Substitute

$p = 1013$ mbar
$T = 273 + 23 = 296$ K

Thus

$$n = 7.243 \times 10^{21} \times 1013/296 = 2.48 \times 10^{22}\ \text{molec. dm}^{-3}$$

and

$$L = n^{-(1/3)} = \left(2.48 \times 10^{22}\right)^{-(1/3)} = 3.43 \times 10^{-8}\ \text{dm molec.}^{-1}.$$

**Exercise 1.21**

A leaktight quartz lightbulb of 10-cm³ volume was operated for a long time at a temperature of 800°C. The pressure of gas in the bulb at filling at 23°C was 140 mbar and then after operation at 800°C and cooling to 23°C was 152 mbar. Assuming that the pressure rise occurred only as a result of the permeation of helium from atmosphere through the walls of the bulb, calculate the mass of helium in the bulb.

**Solution 1.21**

Use Eq. 1.12a written as

$$W_t = [(p_{\text{final}} - p_{\text{initial}})V/(RT)]M$$

and substitute

$$\begin{aligned} p_{\text{final}} &= 152 \text{ mbar} \\ p_{\text{initial}} &= 140 \text{ mbar} \\ V &= 10^{-2} \text{ liters} \\ R &= 83.14 \text{ mbar liters mol}^{-1}\text{ K}^{-1} \text{ (see Table 1.3)} \\ T &= 273 + 23 = 296 \text{ K} \\ M_{\text{He}} &= 4.0 \times 10^{-3} \text{ kg mol}^{-1} \end{aligned}$$

Thus

$$\begin{aligned} W_t &= ((152 - 140) \text{ mbar} \times 10^{-2} \text{ liters}/ \\ &\qquad (83.14 \text{ mbar liters mol}^{-1}\text{ K}^{-1} \times 296 \text{ K})) \times 4.010^{-3} \text{ kg mol}^{-1} \\ &= 1.95 \times 10^{-8} \text{ kg} = 1.95 \times 10^{-2} \text{ mg} \end{aligned}$$

**Exercise 1.22**

Calculate the number of molecules per cubic centimeter at 23°C and $10^{-12}$ mbar.

**Solution 1.22**

Use Eq. 1.8d:

$$n = 7.243 \times 10^{21} p/T \text{ (molec. dm}^{-3})$$

Substitute

$$p = 10^{-12} \text{ mbar}, \qquad T = 273 + 23 = 296 \text{ K}$$

Thus

$$\begin{aligned} n &= 7.243 \times 10^{21} \times (10^{-12} \text{ mbar}/296 \text{ K}) = 2.4 \times 10^{7} \text{ molec. dm}^{-3} \\ &= 2.4 \times 10^{4} \text{ molec. cm}^{-3} \end{aligned}$$

## References

AVS (1970), *Dictionary of Terms for Vacuum Science and Technology, Surface Science, Thin Film Technology, Vacuum Metallurgy, Electronic Materials*, M. S. Kaminsky and J. Lafferty (eds.), American Institute of Physics, New York.

Berman, A. (1985), *Total Pressure Measurements in Vacuum Technology*, Academic Press, New York.

*Compendium of Chemical Terminology*, IUPAC Recommendations (1987), Blackwell Scientific Publications, Oxford, U.K.

CRC (1986–1987), *Handbook of Chemistry and Physics*, 67th ed., R. C. Weast (ed.), CRC Press, Boca Raton, Fla.

ISO/DIS 3529/I (1981). Vacuum technology—vocabulary. Part 1: General terms.

Moss, T. A. (1987), *J. Vac. Sci. Technol.* A5, 5, 2962.

Norton, F. J. (1962), *Transactions of 8th National Vacuum Symposium*, U.S.A., 1961, p. 8, Pergamon Press.

# 2

# Real Gases

## 2.1 *The van der Waals Equation of State*

The behavior of gases near critical temperature and pressure is satisfactorily described by the van der Waals equation of state. Thus, for $n_M = (W_t/M) =$ 1 mol

$$(p + a/V^2)(V - b) = RT \tag{2.1}$$

and for $n_M = (W_t/M) > 1$ mol

$$(p + n_M^2 a/V^2)(V - n_M b) = n_M RT \tag{2.2}$$

where $a$ and $b$ are constants.

The constant $a$ is a measure of intermolecular attraction. The constant $b$ is a measure of the actual volume of the total number of molecules. The values of $b$ increases with increase in molecular diameter $\delta_m$. Thus

$$b = 4N_A(4/3)(\pi\delta_m^3/8) \tag{2.3}$$

The values of the constants $a$ and $b$ can be determined experimentally for each species of gas as a function of the critical temperature $T_c$ and pressure $p_c$ of the gas in question. The values of $T_c$, $p_c$, $a$, and $b$ for various gases and water vapor are listed in Table 1.2 (in Chapter 1).

According to the van der Waals equation of state related to $T_c$, $p_c$, and $V_c$ (critical volume)

$$a = 27R^2T_c^2/(64p_c) \qquad \left[\text{e.g., Pa (cm}^3\text{ mol}^{-1})^2\right] \tag{2.4}$$

$$= 27b^2p_c \tag{2.4a}$$

$$b = RT_c/(8p_c) \qquad (\text{e.g., cm}^3\text{ mol}^{-1}) \tag{2.5}$$

At relatively low pressures, when $V \gg b$, Eq. 2.1 becomes

$$pV = RT(1 + B/V) \tag{2.6}$$

$$= RT + Bp \tag{2.6a}$$

where $B$ is the *second virial coefficient*, expressed as

$$B = b - (a/RT) \qquad (\text{e.g., cm}^3\ \text{mol}^{-1}) \tag{2.7}$$

The quantities by which $a$, $b$, and $B$ are expressed in different systems of units are

| | |
|---|---|
| $a$ | $\text{Pa}\ (\text{cm}^3\ \text{mol}^{-1})^2$ |
| $b, B$ | $(\text{cm}^3\ \text{mol}^{-1})$ |

if $R$ (Pa cm$^3$ mol$^{-1}$ K$^{-1}$), $T_c$ (K), $p_c$ (Pa) in Eq. 2.1,

| | |
|---|---|
| $a$ | $\text{mbar}\ (\text{liters mol}^{-1})^2$ |
| $b, B$ | $(\text{liters mol}^{-1})$ |

if $R$ (mbar liter$^{-1}$ mol$^{-1}$ K$^{-1}$), $T_c$ (K), $p_c$ (mbar) in Eq. 2.1, and

| | |
|---|---|
| $a$ | $\text{atm}\ (\text{cm}^3\ \text{mol}^{-1})^2$ |
| $b, B$ | $(\text{cm}^3\ \text{mol}^{-1})$ |

If $R$ (atm cm$^3$ mol$^{-1}$ K$^{-1}$), $T_c$ (K), $p_c$ (atm) in Eq. 2.1.

The *limiting temperature* $T_B$ to which the ideal gas equation may be applied up to high pressures, also known as *Boyle's point*, is induced by setting in Eq. 2.7 $B = 0$ and $T = T_b$. Thus

$$T_B = a/(bR) \qquad (\text{K}) \tag{2.8}$$

## *Exercises*

**Exercise 2.1**

Calculate the second virial coefficient $B$ at $T = 23°\text{C}$ and the Boyle's point $T_B$ for (a) hydrogen, (b) argon, and (c) xenon. Use the SI system of units.

**Solution 2.1**

For each gas, use Eqs. 2.5, 2.4a, 2.7, and 2.8:

$$b = RT_c/(8p_c)$$
$$a = 27b^2p_c$$
$$B = b - (a/RT)$$
$$T_B = a/(bR)$$
$$T = 273 + 23 = 296\ \text{K}$$
$$R = 8.314\ \text{J mol}^{-1}\ \text{K}^{-1}\ (\text{see Table 1.3})$$
$$= 8.314\ (\text{N m}^{-2})\ \text{m}^3\ \text{mol}^{-1}\ \text{K}^{-1} = 8.314\ \text{Pa cm}^3\ \text{mol}^{-1}\ \text{K}^{-1}$$

1 atm = $1.01 \times 10^5$ Pa (see Table 1.6)

(a) Hydrogen

Substitute

$$T_c = 32.26 \text{ K (see Table 1.2)}$$
$$p_c = 12.8 \text{ atm (see Table 1.2)}$$
$$= 12.8 \text{ atm} \times 1.01 \times 10^5 \text{ Pa atm}^{-1} = 1.29 \times 10^6 \text{ Pa}$$

Thus

$$b_{H_2} = 8.314 \times 10^6 \text{ Pa cm}^3 \text{ mol}^{-1} \text{ K}^{-1} \times 32.26 \text{ K}/8 \times 1.29 \times 10^6 \text{ Pa}$$
$$= 25.99 \text{ cm}^3 \text{ mol}^{-1}$$
$$a_{H_2} = 27 \times 25.99^2 \text{ (cm}^3 \text{ mol}^{-1})^2 \times 1.29 \times 10^6 \text{ Pa}$$
$$= 2.35 \times 10^{10} \text{ Pa (cm}^3 \text{ mol}^{-1})^2$$
$$B_{H_2} = 25.99 \text{ cm}^3 \text{ mol}^{-1} - [2.35 \times 10^{10} \text{ Pa (cm}^3 \text{ mol}^{-1})^2]/$$
$$(8.314 \times 10^6 \text{ Pa cm}^3 \text{ mol}^{-1} \text{ K}^{-1} \times 296 \text{ K})$$
$$= 16.44 \text{ cm}^3 \text{ mol}^{-1}$$
$$T_B = 2.35 \times 10^{10} \text{ Pa cm}^6 \text{ mol}^2/$$
$$(25.99 \text{ cm}^3 \text{ mol}^{-1} \times 8.314 \times 10^6 \text{ Pa cm}^3 \text{ mol}^{-1} \text{ K}^{-1}) = 108.76 \text{ K}$$
$$= -164.44°\text{C}$$

(b) Argon

Substitute

$$T_c = 150.86 \text{ K (see Table 1.2)}$$
$$p_c = 48.0 \text{ atm (see Table 1.2)}$$
$$= 48 \text{ atm} \times 1.01 \times 10^5 \text{ Pa atm}^{-1} = 4.85 \times 10^6 \text{ Pa}$$

Thus

$$b_{Ar} = 8.314 \times 10^6 \text{ Pa cm}^3 \text{ mol}^{-1} \text{ K}^{-1} \times 150.86 \text{ K}/8 \times 4.85 \times 10^6 \text{ Pa}$$
$$= 32.33 \text{ cm}^3 \text{ mol}^{-1}$$
$$a_{Ar} = 27 \times 32.33^2 \text{ (cm}^3 \text{ mol}^{-1})^2 \times 4.85 \times 10^6 \text{ Pa}$$
$$= 1.37 \times 10^{11} \text{ Pa cm}^6 \text{ mol}^{-2}$$
$$B_{Ar} = 32.33 \text{ cm}^3 \text{ mol}^{-1} - [1.37 \times 10^{11} \text{ Pa (cm}^3 \text{ mol}^{-1})^2]/$$
$$(8.314 \times 10^6 \text{ Pa cm}^3 \text{ mol}^{-1} \text{ K}^{-1} \times 296 \text{ K})$$
$$= -23.34 \text{ cm}^3 \text{ mol}^{-1}$$
$$T_B = 1.37 \times 10^{11} \text{ Pa cm}^6 \text{ mol}^{-2}/$$
$$(32.33 \text{ cm}^3 \text{ mol}^{-1} \times 8.314 \times 10^6 \text{ Pa cm}^3 \text{ mol}^{-1} \text{ K}^{-1})$$
$$= 509.68 \text{ K} = 236.52°\text{C}$$

(c) Xenon

Substitute

$$T_c = 289.76 \text{ K (see Table 1.2)}$$
$$p_c = 58 \text{ atm (see Table 1.2)}$$
$$= 58 \text{ atm} \times 1.01 \times 10^5 \text{ Pa atm}^{-1} = 5.86 \times 10^6 \text{ Pa}$$

Thus

$$\begin{aligned}
b_{\mathrm{Xe}} &= 8.314 \times 10^6 \text{ Pa cm}^3 \text{ mol}^{-1} \text{ K}^{-1} \times 289.76 \text{ K}/8 \times 5.86 \times 10^6 \text{ Pa} \\
&= 51.39 \text{ cm}^3 \text{ mol}^{-1} \\
a_{\mathrm{Xe}} &= 27 \times 51.39^2 \text{ (cm}^3 \text{ mol}^{-1})^2 \times 5.86 \times 10^6 \text{ Pa} \\
&= 4.18 \times 10^{11} \text{ Pa cm}^6 \text{ mol}^{-2} \\
B_{\mathrm{Xe}} &= 51.39 \text{ cm}^3 \text{ mol}^{-1} - [4.18 \times 10^{11} \text{ Pa (cm}^3 \text{ mol}^{-1})^2]/ \\
&\quad (8.314 \times 10^6 \text{ Pa cm}^3 \text{ mol}^{-1} \text{ K}^{-1} \times 296 \text{ K}) \\
&= -118.5 \text{ cm}^3 \text{ mol}^{-1} \\
T_{\mathrm{B}} &= 4.18 \times 10^{11} \text{ Pa cm}^6 \text{ mol}^{-2}/ \\
&\quad (51.39 \text{ cm}^3 \text{ mol}^{-1} \times 8.314 \times 10^6 \text{ Pa cm}^3 \text{ mol}^{-1} \text{ K}^{-1}) \\
&= 978.33 \text{ K} = 705.17^\circ\text{C}
\end{aligned}$$

**Exercise 2.2**

One source of systematic uncertainty that affects the calculation of the pressure generated in an expansion calibration system is the deviation from the ideal gas law. Thus, the pressure generated in a single-stage expansion system, derived on assuming that the ideal gas law is obeyed, is (see also Fig. 2A)

$$p = p_1 V_2/(V_2 + V_3)$$

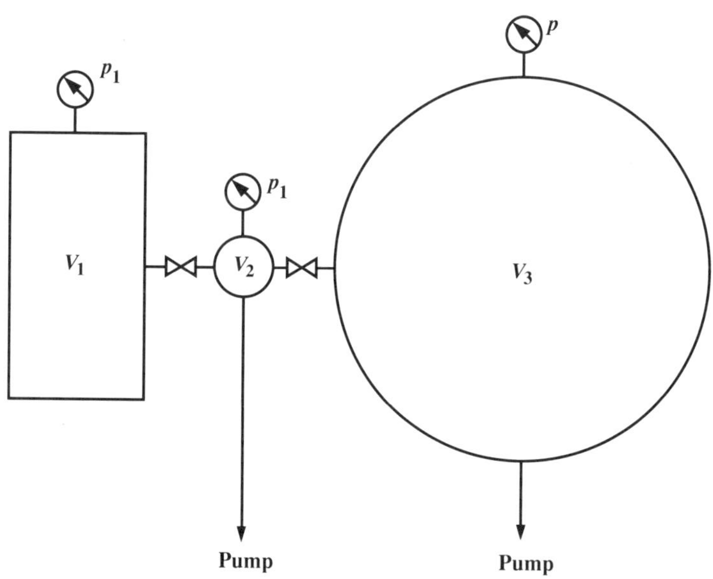


Figure 2A

Determine the derivations of the pressure $p$ calculated from this formula from the pressure generated by (a) hydrogen and (b) xenon in a single stage expansion system where $V_2 = 10\ \text{cm}^3$, $V_3 = 10^4\ \text{cm}^3$, $p_1 = 80\ \text{kPa}$, $T = 23°\text{C}$ (maintained constant during the generation of $p$).

**Solution 2.2**

According to the general gas law (Eq. 1.1), the pressure generated in the expansion calibration system is

(i) $$p = p_1V_2/(V_2 + V_3)$$

The van der Waals equation of state (Eq. 2.6) for low pressures is

(ii) $$pV = RT + Bp$$

Multiply out (ii) by

$$(1/RT)(p_1/p)$$

Thus

$$(1/RT)(p_1/p)pV = (1/RT)(p_1/p)RT + (1/RT)(p_1/p)Bp$$

Add and subtract 1 to the first left-hand side term, then rearrange the terms of the equation. Hence

$$(1/RT)(p_1/p)pV + 1 - 1 = (1/RT)(p_1/p)RT + (1/RT)(p_1/p)Bp$$

$$p_1V/RT + 1 - 1 = p_1/p + Bp_1/RT$$

$$1 - Bp_1/RT = 1 + p_1/p - p_1V/RT$$

Multiply out by $p_1V_2/(V_2 + V_3)$ both terms of this equation (immediately above)

$$[p_1V_2/(V_2 + V_3)](1 - Bp_1/RT)$$

$$= p_1V_2/(V_2 + V_3) \times (1 + p_1/p - p_1V/RT)$$

Substitute $p_1V_2/(V_2 + V_3)$ by $p$, according to (ii). Thus

$$[p_1V_2/(V_2 + V_3)](1 - Bp_1/RT) = p(1 + p_1/p - p_1V/RT)$$

$$(1 + p_1/p - p_1V/RT) = 1 \quad \text{because}$$

$$V/RT = (V_2 + V_3)/RT = 1/p$$

The generated pressure $p$, corrected for the deviation from the ideal gas law, will be

(iii) $$p = p_1V_2/(V_2 + V_3)(1 - Bp_1/RT)$$

Substitute in (iii)

$$\begin{aligned} p_1 &= 80 \text{ kPa} \\ V_1 &= 10 \text{ cm}^3 \\ V_2 &= 10^4 \text{ cm}^3 \\ R &= 8.314 \times 10^6 \text{ Pa cm}^3 \text{ mol}^{-1} \text{ K}^{-1} \text{ (see Table 1.3)} \\ T &= 273 + 23 = 296 \text{ K} \\ B_{H_2} &= 16.44 \text{ cm}^3 \text{ mol}^{-1} \text{ (see Exercise 2.1)} \\ B_{Xe} &= -118.5 \text{ cm}^3 \text{ mol}^{-1} \text{ (see Exercise 2.1)} \end{aligned}$$

and calculate the first term of Eq. (iii):

$$\begin{aligned} p_1V_2/(V_2 + V_3) &= 8.0 \times 10^4 \text{ Pa} \times 10^4 \text{ cm}^3/(10^4 \text{ cm}^3 + 10 \text{ cm}^3) \\ &= 79{,}920 \text{ Pa} \end{aligned}$$

This is the pressure $p$, calculated from Eq. 1.1).

Calculate the correction factor for each gas. Thus

(a) Hydrogen

$$\begin{aligned} 1 - B_{H_2}p_1/RT &= 1 - (16.44 \text{ cm}^3 \text{ mol}^{-1} \times 8.0 \times 10^4 \text{ Pa})/ \\ &\qquad (8.314 \times 10^6 \text{ Pa cm}^3 \text{ mol}^{-1} \text{ K}^{-1} \times 296 \text{ K}) \\ &= 0.99465 \end{aligned}$$

According to (iii) the corrected pressure $p$ would be

$$p = 79{,}920 \text{ Pa} \times 0.99465 = 79{,}942.4 \text{ Pa instead of } 79{,}920 \text{ Pa}$$

calculated from Eq. 1.1.

(b) Xenon

$$\begin{aligned} 1 - B_{Xe}p_1/RT &= 1 - (-118.5 \text{ cm}^3 \text{ mol}^{-1} \times 8.0 \times 10^4 \text{ Pa})/ \\ &\qquad (8.314 \times 10^6 \text{ Pa cm}^3 \text{ mol}^{-1} \text{ K}^{-1} \times 296 \text{ K}) \\ &= 1.00385 \end{aligned}$$

The corrected pressure $p$ would be

$$p = 79{,}920 \text{ Pa} \times 1.00385 = 80{,}277.7 \text{ Pa instead of } 79{,}920 \text{ Pa}$$

calculated from Eq. 1.10.

# 3

# Kinetic Theory of Gases

## 3.1 *Molecular Velocities*

### 3.1.1 The Most Probable Velocity $v_p$

$$v_p = (2RT/M)^{1/2} \qquad (\text{e.g., m s}^{-1}) \tag{3.1}$$

$$= (2kT/m)^{1/2} \tag{3.1a}$$

In the SI system of units

$$v_p = 4.077(T/M)^{1/2} \qquad (\text{m s}^{-1}) \tag{3.1b}$$

if in Eq. 3.1, $R$ (J mol$^{-1}$ K$^{-1}$), $T$ (K), M (kg mol$^{-1}$).

In the CGS system of units

$$v_p = 12{,}895(T/M)^{1/2} \qquad (\text{cm s}^{-1}) \tag{3.1c}$$

if in Eq. 3.1, $R$ (ergs mol$^{-1}$ K$^{-1}$), $T$ (K), $M$ (g mol$^{-1}$).

The significance of the most probable velocity $v_p$ is that more molecules have this velocity than do other molecules.

The values of $v_p$ calculated from Eq. 3.1b for various species of gas at 0°C and 23°C are listed in Table 3.1.

### 3.1.2 The Arithmetic Average Velocity $v_a$

$$v_a = (8RT/\pi M)^{1/2} \qquad (\text{e.g., m s}^{-1}) \tag{3.2}$$

$$= (8kT/\pi m)^{1/2} \tag{3.2a}$$

$$= (8p/(\pi\rho))^{1/2} \qquad (\text{see Eq. 1.14b}) \tag{3.2b}$$

In the SI system of units

$$v_a = 4.602(T/M)^{1/2} \qquad (\text{m s}^{-1}) \tag{3.2c}$$

if in Eq. 3.2, $R$ (J mol$^{-1}$ K$^{-1}$), $T$ (K), M (kg mol$^{-1}$).

**Table 3.1**
The Most Probable Velocity $v_p$ of Molecules of Some Gases at $T$ = 0°C and $T$ = 23°C[a]

| $T$ (K) \ Gas | $v_p$ (m $s^{-1}$) | | | | | | | | | | | | |
|---|---|---|---|---|---|---|---|---|---|---|---|---|---|
| | Monoatomic | | | | | Diatomic | | | | | Polyatomic | | |
| | He | Ne | Ar | Kr | Xe | $H_2$ | $N_2$ | CO | $O_2$ | Air | $NH_3$ | $C_2H_2$ | $CO_2$ |
| 273.16 | 1065 | 474 | 337 | 233 | 186 | 1501 | 418 | 403 | 377 | 396 | 516 | 418 | 312 |
| 296.16 | 1109 | 494 | 351 | 242 | 194 | 1563 | 435 | 419 | 392 | 412 | 538 | 435 | 335 |

[a]Calculated from Eq. 3.1b.

In the CGS system of units

$$v_a = 14{,}554\,(T/M)^{1/2} \qquad (\text{cm s}^{-1}) \tag{3.2d}$$

if in Eq. 3.2, $R$ (ergs mol$^{-1}$ K$^{-1}$), $T$ (K), $M$ (g mol$^{-1}$).

The values of $v_a$ calculated from Eq. 3.2c for various species of gas at 0°C and 23°C are listed in Table 3.2.

The arithmetic average velocity of gases having low molar masses is of the order of kilometers per second (km s$^{-1}$) (see Table 3.2) and therefore is useful in the leak-detection technique.

### 3.1.3 The Mean-Square Velocity $\bar{v}_r^2$

$$\bar{v}_r^2 = (3RT/M) \qquad (\text{e.g., m s}^{-1}) \tag{3.3}$$

$$= (3kT/m) \tag{3.3a}$$

**Table 3.2**
Arithmetic Average Velocity $v_a$ of Molecules of Some Gases at $T$ = 0°C and $T$ = 23°C[a]

| $T$ (K) \ Gas | $v_a$ (m $s^{-1}$) | | | | | | | | | | | | |
|---|---|---|---|---|---|---|---|---|---|---|---|---|---|
| | Monoatomic | | | | | Diatomic | | | | | Polyatomic | | |
| | He | Ne | Ar | Kr | Xe | $H_2$ | $N_2$ | CO | $O_2$ | Air | $NH_3$ | $C_2H_2$ | $CO_2$ |
| 273.16 | 1202 | 535 | 381 | 263 | 210 | 1694 | 472 | 454 | 425 | 447 | 583 | 471 | 363 |
| 296.16 | 1252 | 557 | 396 | 274 | 219 | 1764 | 491 | 473 | 443 | 465 | 607 | 491 | 378 |

[a]Calculated from Eq. 3.2c.

In the SI system of units

$$\bar{v}_r^2 = 24.942(T/M) \qquad (\text{m s}^{-1}) \tag{3.3b}$$

if in Eq. 3.3, $R$ (J mol$^{-1}$ K$^{-1}$), $T$ (K), M (kg mol$^{-1}$).

In the CGS system of units

$$\bar{v}_r^2 = 2.4942 \times 10^8(T/M) \qquad (\text{cm s}^{-1}) \tag{3.3c}$$

if in Eq. 3.3, $R$ (erg mol$^{-1}$ K$^{-1}$), $T$ (K), M (g mol$^{-1}$).

### 3.1.4 The Root-Mean-Square Velocity $v_r$

$$v_r = \left(\bar{v}_r^2\right)^{1/2} \qquad (\text{e.g., m s}^{-1}) \tag{3.4}$$

$$= (3RT/M)^{1/2} \tag{3.4a}$$

$$= (3kT/m)^{1/2} \tag{3.4b}$$

$$= (3p/nm)^{1/2} \qquad (\text{see also Eq. 1.15b}) \tag{3.4c}$$

$$= (3p/\rho)^{1/2} \qquad (\text{see also Eq. 1.14}) \tag{3.4d}$$

In the SI system of units

$$v_r = 4.994(T/M)^{1/2} \qquad (\text{m s}^{-1}) \tag{3.4e}$$

if in Eq. 3.4a, $R$ (J mol$^{-1}$ K$^{-1}$), $T$ (K), M (kg mol$^{-1}$).

In the CGS system of units

$$v_r = 15{,}793(T/M)^{1/2} \qquad (\text{cm s}^{-1}) \tag{3.4f}$$

if in Eq. 3.4a, $R$ (erg mol$^{-1}$ K$^{-1}$), $T$ (K), M (g mol$^{-1}$)

*The root-mean-square velocities of various species of gas, at the same temperature, are inversely proportional to the square roots of their molecular mass m, or molar mass* M. Thus

$$v_{r1}/v_{r2} = (M_2/M_1)^{1/2} = (m_2/m_1)^{1/2} \tag{3.4g}$$

Table 3.3 lists the values of $v_r$ calculated from Eq. 3.4e for various species of gas at 0°C and 23°C.

**Table 3.3**
The Root-Mean-Square Velocity $v_r$ of Molecules of Some Gases at $T$ = 0°C and 23°C[a]

| $T$ (K) \ Gas | $v_r$ (m s$^{-1}$) | | | | | | | | | | | | |
|---|---|---|---|---|---|---|---|---|---|---|---|---|---|
| | Monoatomic | | | | | Diatomic | | | | | Polyatomic | | |
| | He | Ne | Ar | Kr | Xe | $H_2$ | $N_2$ | CO | $O_2$ | Air | $NH_3$ | $C_2H_2$ | $CO_2$ |
| 273.16 | 1305 | 581 | 413 | 285 | 228 | 1838 | 512 | 493 | 461 | 485 | 632 | 511 | 393 |
| 296.16 | 1358 | 605 | 430 | 297 | 237 | 1914 | 533 | 514 | 480 | 505 | 659 | 533 | 410 |

[a]Calculated from Eq. 3.4e.

## 3.2 Relationships between Molecular Velocities

$$v_p = 0.866 v_a \tag{3.5}$$

$$v_p = 0.816 v_r \tag{3.6}$$

$$v_r = 1.085 v_a \tag{3.7}$$

(*Note*: The values of the ratios $v_p/v_a$, $v_p/v_r$, and $v_r/v_a$ have been rounded down to the third digit after the decimal point.)

## 3.3 The Maxwell–Boltzmann Law of Distribution of Molecular Velocities

$$f_v = (1/n)(dn/dv) \tag{3.8}$$

$$= (4/\pi^{1/2})(m/2kT)^{3/2} v^2 \exp(-mv^2/2kT) \tag{3.8a}$$

where $f_v$ is the fractional number of molecules in the range of velocities between $v$ and $v + dv$ per unit velocity range.

Figure 3.1 shows the Maxwell–Boltzmann distribution curves for the same gas at different temperatures (in kelvins). The total area under either curve estimates the total number of molecules. Hence, if the curves refer to the same number of molecules, the areas must be the same.

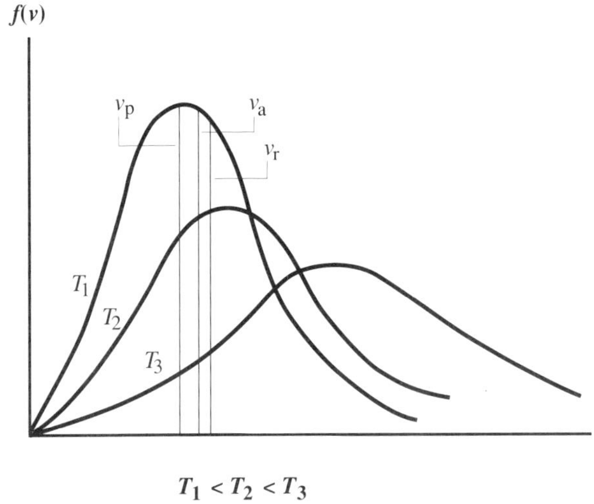


**Figure 3.1** Maxwell–Boltzmann distribution curves for the same gas at various temperatures.

## 3.4 *Kinetic Energy of Molecules*

### 3.4.1 The Average Translational Energy $\bar{E}$ per Molecule in Random Motion

$$\bar{E} = (1/2)(m\bar{v}^{1/2}) \qquad (\text{J molec.}^{-1}) \tag{3.9}$$

$$= (3/2)(kT) \tag{3.9a}$$

*The average translational energy $\bar{E}$ per molecule in random motion is the same for all gases and is proportional to the absolute temperature.*

Figure 3.2 represents a plot of Eq. 3.9 for different temperatures.

### 3.4.2 The Total Translational Energy E of Molecules in Random Motion

$$E = (3/2)pV \qquad (\text{J}) \tag{3.10}$$

$$= (3/2)(W_t/M)RT \tag{3.11}$$

$$= (1/2)W_t\bar{v}^2 \tag{3.12}$$

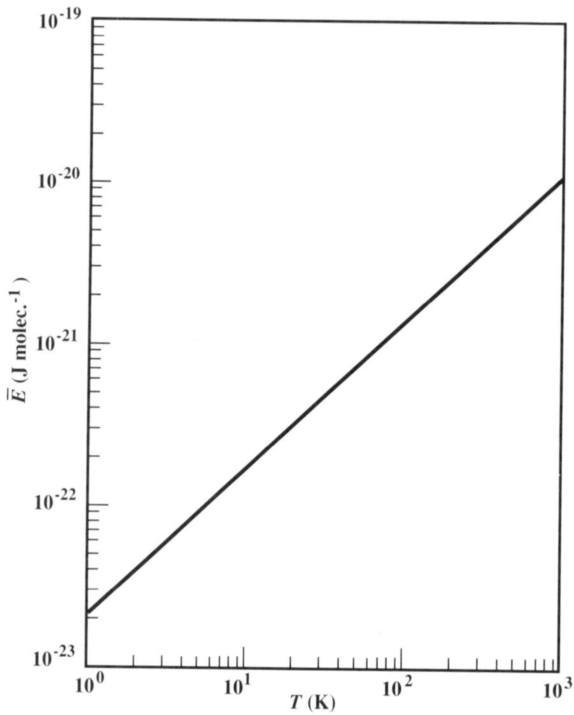


**Figure 3.2** Average translational energy $\overline{E}$ per molecule in random motion in a gas as a function of pressure.

## 3.5 *Quantitative Relationships between Molecules and Areas*

### 3.5.1 The Rate $\phi$ at Which Molecules at Steady State Strike a Unit Area per Unit Time

$$\phi = (1/4)nv_a \qquad (\text{e.g., molec. cm}^{-2}\ \text{s}^{-1}) \tag{3.13}$$

The number of molecules that strike an element of surface perpendicular to their direction of movement, per unit time, expressed as a function of pressure in different systems of units is

$$\phi = 8.333 \times 10^{22} p/(TM)^{1/2} \qquad (\text{molec. m}^{-2}\ \text{s}^{-1}) \tag{3.13a}$$

with $p$ (Pa), $T$ (K), and M (kg mol$^{-1}$) (i.e, where $p$ is in pascals; $T$, in Kelvins; and M, in kilograms per mole),

$$\phi = 2.635 \times 10^{22} p/(TM)^{1/2} \qquad (\text{molec. cm}^{-2}\ \text{s}^{-1}) \tag{3.13b}$$

**Table 3.4**
The Rate $\phi$ at Which Molecules at Steady State Strike a Unit Area per Unit Time for Some Gases at $T = 0°C$ as a Function of Pressure (mbar)[a]

| $p$ (mbar) | $\phi$ (molec. $cm^{-2}\ s^{-1}$) He | Ne | Air | $CO_2$ | Kr | Xe |
|---|---|---|---|---|---|---|
| $1.0 \times 10^{3}$ | $7.97 \times 10^{23}$ | $3.55 \times 10^{23}$ | $2.96 \times 10^{23}$ | $2.40 \times 10^{23}$ | $1.74 \times 10^{23}$ | $1.39 \times 10^{23}$ |
| 1.0 | $7.97 \times 10^{20}$ | $3.56 \times 10^{20}$ | $2.96 \times 10^{20}$ | $2.40 \times 10^{20}$ | $1.74 \times 10^{20}$ | $1.39 \times 10^{20}$ |
| $1.0 \times 10^{-3}$ | $7.97 \times 10^{17}$ | $3.55 \times 10^{17}$ | $2.96 \times 10^{17}$ | $2.40 \times 10^{17}$ | $1.74 \times 10^{17}$ | $1.39 \times 10^{17}$ |
| $1.0 \times 10^{-6}$ | $7.97 \times 10^{14}$ | $3.55 \times 10^{14}$ | $2.96 \times 10^{14}$ | $2.40 \times 10^{14}$ | $1.74 \times 10^{14}$ | $1.39 \times 10^{14}$ |
| $1.0 \times 10^{-9}$ | $7.97 \times 10^{11}$ | $3.55 \times 10^{11}$ | $2.96 \times 10^{11}$ | $2.40 \times 10^{11}$ | $1.74 \times 10^{11}$ | $1.39 \times 10^{11}$ |
| $1.0 \times 10^{-12}$ | $7.97 \times 10^{8}$ | $3.55 \times 10^{8}$ | $2.96 \times 10^{8}$ | $2.40 \times 10^{8}$ | $1.74 \times 10^{8}$ | $1.39 \times 10^{8}$ |
| $1.0 \times 10^{-15}$ | $7.97 \times 10^{5}$ | $3.55 \times 10^{5}$ | $2.96 \times 10^{5}$ | $2.40 \times 10^{5}$ | $1.74 \times 10^{5}$ | $1.39 \times 10^{5}$ |

[a]Calculated from Eq. 3.13b.

with $p$ (mbar), $T$ (K), and $M$ (g $mol^{-1}$), and

$$\phi = 3.513 \times 10^{22} p/(TM)^{1/2} \qquad (\text{molec. cm}^{-2}\ \text{s}^{-1}) \qquad (3.13c)$$

with $p$ (Torr), $T$ (K), and $M$ (g $mol^{-1}$).

For air at $T = 23°C$ and $p$ (mbar)

$$\phi = 2.84 \times 10^{20} p \qquad (\text{molec. cm}^{-2}\ \text{s}^{-1}) \qquad (3.13d)$$

Equation 3.13 holds true if the scattering of molecules at the surface they strike follows the cosine law (Edwards, 1978).

The values of $\phi$ for various gases at $T = 0°C$ and $T = 23°C$ as a function of pressure (mbar) are listed in Tables 3.4 and 3.5.

**Table 3.5**
The Rate $\phi$ at Which Molecules at Steady State Strike a Unit Area per Unit Time for Some Gases at $T = 23°C$ as a Function of Pressure (mbar)[a]

| $p$ (mbar) | $\phi$ (molec. $cm^{-2}\ s^{-1}$) He | Ne | Air | $CO_2$ | Kr | Xe |
|---|---|---|---|---|---|---|
| $1.0 \times 10^{3}$ | $7.65 \times 10^{23}$ | $3.41 \times 10^{23}$ | $2.84 \times 10^{23}$ | $2.31 \times 10^{23}$ | $1.67 \times 10^{23}$ | $1.33 \times 10^{23}$ |
| 1.0 | $7.65 \times 10^{20}$ | $3.41 \times 10^{20}$ | $2.84 \times 10^{20}$ | $2.31 \times 10^{20}$ | $1.67 \times 10^{20}$ | $1.33 \times 10^{20}$ |
| $1.0 \times 10^{-3}$ | $7.65 \times 10^{17}$ | $3.41 \times 10^{17}$ | $2.84 \times 10^{17}$ | $2.31 \times 10^{17}$ | $1.67 \times 10^{17}$ | $1.33 \times 10^{17}$ |
| $1.0 \times 10^{-6}$ | $7.65 \times 10^{14}$ | $3.41 \times 10^{14}$ | $2.84 \times 10^{14}$ | $2.31 \times 10^{14}$ | $1.67 \times 10^{14}$ | $1.33 \times 10^{14}$ |
| $1.0 \times 10^{-9}$ | $7.65 \times 10^{11}$ | $3.41 \times 10^{11}$ | $2.84 \times 10^{11}$ | $2.31 \times 10^{11}$ | $1.67 \times 10^{11}$ | $1.33 \times 10^{11}$ |
| $1.0 \times 10^{-12}$ | $7.65 \times 10^{8}$ | $3.41 \times 10^{8}$ | $2.84 \times 10^{8}$ | $2.31 \times 10^{8}$ | $1.67 \times 10^{8}$ | $1.33 \times 10^{8}$ |
| $1.0 \times 10^{-15}$ | $7.65 \times 10^{5}$ | $3.41 \times 10^{5}$ | $2.84 \times 10^{5}$ | $2.31 \times 10^{5}$ | $1.67 \times 10^{5}$ | $1.33 \times 10^{5}$ |

[a]Calculated from Eq. 3.13b.

### 3.5.2 The Mass of Gas G Incident on Unit Area per Unit Time

$$G = m\phi \qquad (\text{e.g., kg m}^{-2}\,\text{s}^{-1}) \tag{3.14}$$

$$= mnv_a/4 \tag{3.14a}$$

$$= \rho v_a/4 \tag{3.14b}$$

The expression of G as a function of pressure in different systems of units is

$$G = 1.383 \times 10^{-1} p(M/T)^{1/2} \qquad (\text{kg m}^{-2}\,\text{s}^{-1}) \tag{3.14c}$$

with $p$ (Pa), $T$ (K), and M (kg mol$^{-1}$),

$$G = 4.376 \times 10^{-2} p(M/T)^{1/2} \qquad (\text{g cm}^{-2}\,\text{s}^{-1}) \tag{3.14d}$$

with $p$ (mbar), $T$ (K), and M (g mol$^{-1}$), and

$$G = 5.834 \times 10^{-2} p(M/T)^{1/2} \qquad (\text{g cm}^{-2}\,\text{s}^{-1}) \tag{3.14e}$$

with $p$ (Torr), $T$ (K), and M (g mol$^{-1}$).

For air at $T$ = 23°C and $p$ (mbar)

$$G = 1.639 \times 10^{-2} p \tag{3.14f}$$

### 3.5.3 Effusion Rate of Molecules $q_e$ through a Small Orifice of Area A

$$q_e = \phi A \qquad (\text{molec. s}^{-1}) \tag{3.15}$$

$$= (1/4) n v_a A \tag{3.15a}$$

The expression of $q_e$ as a function of pressure in different systems of units is

$$q_e = 8.333 \times 10^{22} p/(TM)^{1/2} \qquad (\text{molec. s}^{-1}) \tag{3.15b}$$

with $p$ (Pa), $T$ (K), and M (kg mol$^{-1}$),

$$q_e = 2.635 \times 10^{22} p/(TM)^{1/2} \qquad (\text{molec. s}^{-1}) \tag{3.15c}$$

with $p$ (mbar), $T$ (K), and M (g mol$^{-1}$), and

$$q_e = 3.513 \times 10^{22} p/(TM)^{1/2} \qquad (\text{molec. s}^{-1}) \tag{3.15d}$$

with $p$ (Torr), $T$ (K), and M (g mol$^{-1}$).

Equation 3.15 holds true if, and only if, the orifice separates two regions of "high" and "zero" gas density, respectively, the orifice is cut in a flat plate, the thickness of the plate is vanishingly small in the region of the orifice, and the mean free path of molecules $\lambda$ (see Section 3.8) is very small compared to the diameter of the orifice.

### 3.5.4 The Volume of Gas $\dot{B}$ Impinging on Unit Area per Unit Time

$$\dot{B} = \phi/n \qquad (\text{cm}^3\ \text{cm}^{-2}\ \text{s}^{-1}) \tag{3.16}$$

$$\dot{B} = v_a/4 \tag{3.16a}$$

$$= 3639(T/M)^{1/2} \qquad (\text{cm}^3\ \text{cm}^{-2}\ \text{s}^{-1}) \tag{3.16b}$$

with $T$ (K) and M (g mol$^{-1}$).

*The volume of gas $\dot{B}$ impinging on unit area per unit time is constant at all pressures and varies only with the square of the ratio of the absolute temperature of the gas to its molar mass.*

### 3.5.5 The Volume of Gas $B_o$ Escaping per Unit Time through an Orifice of Area A

$$B_o = q_e/n \qquad (\text{e.g., cm}^3\ \text{s}^{-1}) \tag{3.17}$$

$$= Av_a/4 \tag{3.17a}$$

$$= 3.639A(T/M)^{1/2} \qquad (\text{liters s}^{-1}) \tag{3.17b}$$

with A (cm$^2$), $T$ (K), and M (g mol$^{-1}$) and M (g mol$^{-1}$).

For air at $T$ = 23°C

$$B_o = 11.6A \qquad (\text{liters s}^{-1}) \tag{3.17c}$$

## 3.6 *Methods for Estimating the Molecular Diameter $\delta_m$*

Either of the following methods can be used to estimate the diameter $\delta_m$ of the molecule.

### 3.6.1 Use of the Constant $b$ in the van der Waals Equation of State

$$\delta_m = (7.929 \times 10^{-25} b)^{1/3} \qquad (\text{cm}) \tag{3.18}$$

Note that in Eq. 2.1 the molecules are considered to be rigid elastic spheres.

The values of the constant $b$ for various gases and water vapor are listed in Table 1.2.

Table 3.6

Apparent Diameter $\delta_m$ (with values $10^{-10}$ m) of Molecules of Some Gases[a]

| Monoatomic | | | | | Diatomic | | | | | Polyatomic |
|---|---|---|---|---|---|---|---|---|---|---|
| He | Ne | Ar | Kr | Xe | $H_2$ | $N_2$ | CO | $O_2$ | Air | $CO_2$ |
| 2.20 | 2.55 | 3.69 | 4.27 | 4.87 | 2.68 | 3.78 | 3.79 | 3.65 | 3.76 | 4.66 |

[a]Adapted from Roth (1982), p. 46.

### 3.6.2 Use of the Density ρ of the Solid or Liquid

Assuming that the molecules are closely packed as in a face-centered cubic lattice (Dushman, 1966 p. 38), we find

$$\delta_m = 1.329 \times 10^{-8} (M/\rho)^{1/3} \quad \text{(cm)} \qquad (3.19)$$

with M (g $mol^{-1}$) and $\rho$ (g $dm^{-3}$).

Note that the density of gas $\rho$ depends on temperature and pressure (see Section 1.15).

### 3.6.3 Measurement of the Coefficient of Viscosity η (Eq. 3.27)

$$\delta_m = \left(2.107 \times 10^{-9}(mT)^{1/2}/\eta\right)^{1/2} \quad \text{(cm)} \qquad (3.20)$$

where $m$ (g), $T$ (K), and $\eta$ (dyn s $cm^{-2}$ or poise).

For the experimental determination of the coefficient of viscosity $\eta$ by an accurate method such as the rotation cylinder, see Present (1958, p. 185).

The apparent diameters $\delta_m$ of different gases are listed in Table 3.6.

### 3.6.4 Use of the Coefficient of Interdiffusion of Gases $D_{1,2}$

The expression of the average molecular diameter $\delta_{m1,2}$ induced from Eq. 3.30 for gases consisting of molecules of different diameters $\delta_{m1}$ and $\delta_{m2}$ is

$$\delta_{m1,2}^2 = \left(v_{a1}^2 + v_{a2}^2\right)^{1/2}/3\pi(n_1 + n_2)D_{1,2} \quad (cm^2) \qquad (3.21)$$

where $\delta_{m1,2} = (\delta_{m1} + \delta_{m2})/2$ (cm) is the average molecular diameter, $v_{a1}$, $v_{a2}$ (cm $s^{-1}$), $n_1$, $n_2$ (molec. $cm^{-3}$), and $D_{1,2}$ ($cm^2$ $s^{-1}$).

The value of the average molecular diameter $\delta_{m1,2}$ is listed in Table 3.13.

## *3.7 The Time $t_m$ to Form a Monolayer*

Assume that the rate $\phi$ at which molecules at steady state strike unit area per unit time is $\phi = (1/4)nv_a$ (molec. $cm^{-2}$ $s^{-1}$) (Eq. 3.13) and the number of molecules necessary to form a monolayer is $\phi_m = 1/\delta_m^2$ ($cm^{-2}$); then

$$t_m = \phi_m/\phi \qquad (s) \tag{3.22}$$

$$= 4/\left(nv_a\delta_m^2\right) \tag{3.22a}$$

The time to form a monolayer $t_m$ is usually expressed as a function of pressure. Thus, in different systems of units

$$t_m = 1.199 \times 10^{-23}\,(TM)^{1/2}/\left(p\delta_m^2\right) \qquad (s) \tag{3.22b}$$

with $T$ (K), M (kg $mol^{-1}$), $p$ (Pa), and $\delta_m$ (m),

$$t_m = 3.794 \times 10^{-23}\,(TM)^{1/2}/\left(p\delta_m^2\right) \qquad (s) \tag{3.22c}$$

with $T$ (K), M (g $mol^{-1}$), $p$ (mbar), and $\delta_m$ (cm), and

$$t_m = 2.847 \times 10^{-23}(TM)^{1/2}/\left(p\delta_m^2\right) \qquad (s) \tag{3.22d}$$

with $T$ (K), M (g $mol^{-1}$), and $\delta_m$ (cm).

*Equation 3.22 holds true only when the surface on which the molecules are adsorbed, is clean and the sticking coefficient is unity.*

For air at $T$ = 23°C and $p$ (mbar)

$$t_m = 2.49 \times 10^{-6}/p \qquad (s) \tag{3.22e}$$

**Table 3.7**

The Time $t_m$ to Form a Monolayer for Some Gases at $T$ = 0°C as a Function of Pressure (mbar)[a]

| $P$ (mbar) | $t_m$ (s) $CO_2$ | Air | He | Xe | Kr | Ne |
|---|---|---|---|---|---|---|
| $1.0 \times 10^3$ | $1.91 \times 10^{-9}$ | $2.39 \times 10^{-9}$ | $2.59 \times 10^{-9}$ | $3.03 \times 10^{-9}$ | $3.15 \times 10^{-9}$ | $4.33 \times 10^{-9}$ |
| 1.0 | $1.91 \times 10^{-6}$ | $2.39 \times 10^{-6}$ | $2.59 \times 10^{-6}$ | $3.03 \times 10^{-6}$ | $3.15 \times 10^{-6}$ | $4.33 \times 10^{-6}$ |
| $1.0 \times 10^{-3}$ | $1.91 \times 10^{-3}$ | $2.39 \times 10^{-3}$ | $2.59 \times 10^{-3}$ | $3.03 \times 10^{-3}$ | $3.15 \times 10^{-3}$ | $4.33 \times 10^{-3}$ |
| $1.0 \times 10^{-6}$ | 1.91 | 2.39 | 2.59 | 3.03 | 3.15 | 4.33 |
| $1.0 \times 10^{-9}$ | $1.91 \times 10^3$ | $2.39 \times 10^3$ | $2.59 \times 10^3$ | $3.03 \times 10^3$ | $3.15 \times 10^3$ | $4.33 \times 10^3$ |
| $1.0 \times 10^{-12}$ | $1.91 \times 10^6$ | $2.39 \times 10^6$ | $2.59 \times 10^6$ | $3.03 \times 10^6$ | $3.15 \times 10^6$ | $4.33 \times 10^6$ |
| $1.0 \times 10^{-15}$ | $1.91 \times 10^9$ | $2.39 \times 10^9$ | $2.59 \times 10^6$ | $3.03 \times 10^9$ | $3.15 \times 10^9$ | $4.33 \times 10^9$ |

[a]Calculated from Eq. 3.22c.

**Table 3.8**

The Time $t_m$ to Form a Monolayer for Some Gases at $T = 23°C$ as a Function of Pressure (mbar)[a]

| $p$ | $t_m$ (s) | | | | | |
|---|---|---|---|---|---|---|
| (mbar) | $CO_2$ | Air | He | Xe | Kr | Ne |
| $1.0 \times 10^3$ | $1.99 \times 10^{-9}$ | $2.49 \times 10^{-9}$ | $2.70 \times 10^{-9}$ | $3.15 \times 10^{-9}$ | $3.28 \times 10^{-9}$ | $4.51 \times 10^{-9}$ |
| 1.0 | $1.99 \times 10^{-6}$ | $2.49 \times 10^{-6}$ | $2.70 \times 10^{-6}$ | $3.15 \times 10^{-6}$ | $3.28 \times 10^{-6}$ | $4.51 \times 10^{-6}$ |
| $1.0 \times 10^{-3}$ | $1.99 \times 10^{-3}$ | $2.49 \times 10^{-3}$ | $2.70 \times 10^{-3}$ | $3.15 \times 10^{-3}$ | $3.28 \times 10^{-3}$ | $4.51 \times 10^{-3}$ |
| $1.0 \times 10^{-6}$ | 1.99 | 2.49 | 2.70 | 3.15 | 3.28 | 4.51 |
| $1.0 \times 10^{-9}$ | $1.99 \times 10^3$ | $2.49 \times 10^3$ | $2.70 \times 10^3$ | $3.15 \times 10^3$ | $3.28 \times 10^3$ | $4.51 \times 10^3$ |
| $1.0 \times 10^{-12}$ | $1.99 \times 10^6$ | $2.49 \times 10^6$ | $2.70 \times 10^6$ | $3.15 \times 10^6$ | $3.28 \times 10^6$ | $4.51 \times 10^6$ |
| $1.0 \times 10^{-12}$ | $1.99 \times 10^9$ | $2.49 \times 10^9$ | $2.70 \times 10^9$ | $3.15 \times 10^9$ | $3.28 \times 10^9$ | $4.51 \times 10^9$ |

[a]Calculated from Eq. 3.22c.

Tables 3.7 and 3.8 list the time to form a monolayer at $T = 0°C$ and $T = 23°C$ and pressures ranging from $p = 10^{-3}$ mbar to $p = 10^{-15}$ mbar.

Assuming (Dushman, 1966, p. 420) that the molecules are closely packed and are spheres of diameter $\delta_m$, then the number of molecules $n_m$ adsorbed per square centimeter to form a monolayer is

$$n_m = 1/(1.035\pi\delta_m^2/4) \qquad (\text{molec. cm}^{-2}) \tag{3.22f}$$

## 3.8 The Mean Free Path λ of Molecules in Gas

The distance covered by molecules between successive collisions in gas is given in Eqs. 3.23.

### 3.8.1 Gas Consisting of Molecules of Same Diameter $\delta_m$

$$\lambda = 1/(2^{1/2}\pi n\delta_m^2) \qquad (\text{e.g., m}) \tag{3.23}$$

$$\lambda = kT/(2^{1/2}\pi\delta_m^2 p) \tag{3.23a}$$

$$\lambda = 2\eta/\rho v_a \tag{3.23b}$$

(For the expression of $\eta$, see Eq. 3.27.)

The mean free path $\lambda$ expressed as a function of pressure $p$ in different system of units is

$$\lambda = 3.107 \times 10^{-24} T/(p\delta_m^2) \qquad (\text{m}) \tag{3.23c}$$

**Table 3.9**
The Mean Free Path λ of Some Gases at $T = 0°C$ as a Function of Pressure $P$ (mbar)[a]

| $P$ (mbar) | λ (cm) | | | | | |
|---|---|---|---|---|---|---|
| | He | Ne | Air | Kr | $CO_2$ | Xe |
| $1.0 \times 10^{3}$ | $1.75 \times 10^{-5}$ | $1.31 \times 10^{-5}$ | $6.00 \times 10^{-6}$ | $4.65 \times 10^{-6}$ | $3.91 \times 10^{-6}$ | $3.58 \times 10^{-6}$ |
| 1.0 | $1.75 \times 10^{-2}$ | $1.31 \times 10^{-2}$ | $6.00 \times 10^{-3}$ | $4.65 \times 10^{-3}$ | $3.91 \times 10^{-3}$ | $3.58 \times 10^{-3}$ |
| $1.0 \times 10^{-3}$ | $1.75 \times 10$ | $1.31 \times 10$ | 6.00 | 4.65 | 3.91 | 3.58 |
| $1.0 \times 10^{-6}$ | $1.75 \times 10^{4}$ | $1.31 \times 10^{4}$ | $6.00 \times 10^{3}$ | $4.65 \times 10^{3}$ | $3.91 \times 10^{3}$ | $3.58 \times 10^{3}$ |
| $1.0 \times 10^{-9}$ | $1.75 \times 10^{7}$ | $1.31 \times 10^{7}$ | $6.00 \times 10^{6}$ | $4.65 \times 10^{6}$ | $3.91 \times 10^{6}$ | $3.58 \times 10^{6}$ |
| $1.0 \times 10^{-12}$ | $1.75 \times 10^{10}$ | $1.31 \times 10^{10}$ | $6.00 \times 10^{9}$ | $4.65 \times 10^{9}$ | $3.91 \times 10^{9}$ | $3.58 \times 10^{9}$ |
| $1.0 \times 10^{-15}$ | $1.75 \times 10^{13}$ | $1.31 \times 10^{13}$ | $6.00 \times 10^{12}$ | $4.65 \times 10^{12}$ | $3.91 \times 10^{12}$ | $3.58 \times 10^{12}$ |

[a]Calculated from Eq. 3.23d.

with $T$ (K), $p$ (Pa), and $\delta_m$ (m),

$$\lambda = 3.107 \times 10^{-20} T/(p\delta_m^2) \qquad \text{(cm)} \qquad (3.23\text{d})$$

with $T$ (K), $p$ (mbar), and $\delta_m$ (cm), and

$$\lambda = 2.330 \times 10^{-20} T/(p\delta_m^2) \qquad \text{(cm)} \qquad (3.23\text{e})$$

with $T$ (K), $p$ (Torr), and $\delta_m$ (cm).

**Table 3.10**
The Mean Free Path λ of Some Gases at $T = 23°C$ as a Function of Pressure $P$ (mbar)[a]

| $P$ (mbar) | λ (cm) | | | | | |
|---|---|---|---|---|---|---|
| | He | Ne | Air | Kr | $CO_2$ | Xe |
| $1.0 \times 10^{3}$ | $1.90 \times 10^{-5}$ | $1.42 \times 10^{-5}$ | $6.51 \times 10^{-6}$ | $5.05 \times 10^{-6}$ | $4.24 \times 10^{-6}$ | $3.88 \times 10^{-6}$ |
| 1.0 | $1.90 \times 10^{-2}$ | $1.42 \times 10^{-2}$ | $6.51 \times 10^{-3}$ | $5.05 \times 10^{-3}$ | $4.24 \times 10^{-3}$ | $3.88 \times 10^{-3}$ |
| $1.0 \times 10^{-3}$ | $1.90 \times 10$ | $1.42 \times 10$ | 6.51 | 5.05 | 4.24 | 3.88 |
| $1.0 \times 10^{-6}$ | $1.90 \times 10^{4}$ | $1.42 \times 10^{4}$ | $6.51 \times 10^{3}$ | $5.05 \times 10^{3}$ | $4.24 \times 10^{3}$ | $3.88 \times 10^{3}$ |
| $1.0 \times 10^{-9}$ | $1.90 \times 10^{7}$ | $1.42 \times 10^{7}$ | $6.51 \times 10^{6}$ | $5.05 \times 10^{6}$ | $4.24 \times 10^{6}$ | $3.88 \times 10^{6}$ |
| $1.0 \times 10^{-12}$ | $1.90 \times 10^{10}$ | $1.42 \times 10^{10}$ | $6.51 \times 10^{9}$ | $5.05 \times 10^{9}$ | $4.24 \times 10^{9}$ | $3.88 \times 10^{9}$ |
| $1.0 \times 10^{-15}$ | $1.90 \times 10^{13}$ | $1.42 \times 10^{13}$ | $6.51 \times 10^{12}$ | $5.05 \times 10^{12}$ | $4.24 \times 10^{12}$ | $3.88 \times 10^{12}$ |

[a]Calculated from Eq. 3.23d.

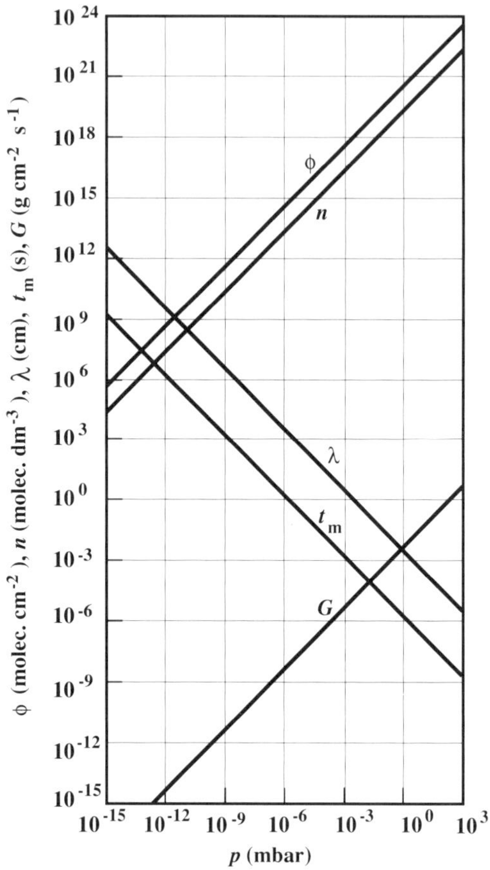


**Figure 3.3** Variation with pressure $p$ (mbar) of $n$ (molec. dm$^3$) for all gases at $T = 23°C$ (calculated from Eq. 1.8d), $\phi$ (molec. cm$^{-2}$) for $N_2$ at $T = 23°C$ (calculated from Eq. 3.13b), G (g cm$^{-2}$ s$^{-1}$) for $N_2$ at $T = 23°C$ (calculated from Eq. 3.14d), $t_m$ (s) for $N_2$ at T = 23°C (calculated from Eq. 3.22c), and $\lambda$ (cm) for $N_2$ at $T = 23°C$ (calculated from Eq. 3.23d).

For air at $T = 23°C$ (assuming that all the molecules that compose the gas have the same diameter $\delta_m$)

$$\lambda = 6.5 \times 10^{-3}/p_{Pa} \qquad (m) \tag{3.23f}$$

$$\lambda = 4.88 \times 10^{-3}/p_{Torr} \qquad (cm) \tag{3.23g}$$

*The mean free path λ of molecules is inversely proportional to the gas density. Therefore, at a given temperature, λ is inversely proportional to the gas pressure.*

Tables 3.9 and 3.10 list the mean free path $\lambda$ for some gases at $T = 0°C$ and $T = 23°C$ calculated from Eq. 3.23d.

Figure 3.3 illustrates the variation with pressure $p$ (mbar) of the number density of molecules $n$ *for all gases* at $T = 23°C$, the rate $\phi$ at which molecules at steady state strike unit area per unit time, the mass of gas G incident on unit area per unit time, the time $t_m$ to form a monolayer, and the mean free path of molecules $\lambda$, all calculated for $N_2$ at $T = 23°C$.

At very high temperatures $T_\infty$ and constant number density of molecules ($n$ = const. in Eq. 3.23) the mean free path of molecules is

$$\lambda_T = \lambda_\infty/(1 + C_s/T) \qquad (e.g., cm) \tag{3.23h}$$

where $C_s$ is the Sutherland constant.

The values of Sutherland's constant $C_s$ for several gases and water vapor are listed in Table 3.11.

### 3.8.2 Mixture of Gases Consisting of Molecules of Various Diameters

$$\lambda_{1,2} = 4kT/\left(\pi(\delta_{m1} + \delta_{m2})^2 p_2(1 + M_1/M_2)^{1/2}\right) \qquad (cm) \tag{3.23i}$$

$$\lambda_{1,2} = 4 \times 2^{1/2}\lambda_2\left(1 + (\lambda_2/\lambda_1)^{1/2}\right)^{-2}\left[1 + (M_1/M_2)^{-1/2}\right] \tag{3.23j}$$

where $\delta_{m1}$ and $\delta_{m2}$ are the molecular diameters of the different gases.

**Table 3.11**

Sutherland's Constant $C_s$ (K) for Some Gases and Vapor[a]

| Monoatomic | | | | | | Diatomic | | | | | Polyatomic | |
|---|---|---|---|---|---|---|---|---|---|---|---|---|
| He | Ne | Ar | Kr | Xe | Hg | $H_2$ | $N_2$ | Air[b] | CO | $O_2$ | $CO_2$ | $H_2O$ |
| 79 | 56 | 169 | 142 | 252 | 942 | 76 | 112 | 112 | 100 | 132 | 273 | 600 |

[a]After Diels and Jaeckel (1958), p. 8.
[b]After Dushman (1966) p. 32.

## 3.9 *The Mean Free Path of Electrons* $\boldsymbol{\lambda}_e$ *in Gas*

$$\lambda_e = 4 \times 2^{1/2}\lambda \qquad \text{(cm)} \tag{3.24}$$

$$= 1.273/(\delta_m^2 n) \tag{3.24a}$$

where $\lambda$ (cm) is the mean free path of molecules.

## 3.10 *The Mean Free Path of Ions* $\boldsymbol{\lambda}_i$ *in Gas*

$$\lambda_i = 2^{1/2}\lambda \qquad \text{(cm)} \tag{3.25}$$

where $\lambda$ (cm) is the mean free path of molecules.

## 3.11 *Transport Phenomena in the Viscous State*

### 3.11.1 The Knudsen Number $K'$

$$K' = \lambda/d \qquad \text{(dimensionless)} \tag{3.26}$$

where $\lambda$ is the mean free path of molecules and $d$ is a characteristic dimension of the vessel through which gas flows (e.g., the diameter for cylindrical pipes). Note that $\lambda$ and $d$ must be expressed in the same units (e.g., m). The Knudsen number (Knudsen, 1910) permits the determination of the gas flow regime (see Section 4.6.1.c).

*The viscous flow of gas is characterized by intermolecular collisions that impart to the gas the pattern of a viscous fluid.*

### 3.11.2 The Coefficient of Viscosity $\boldsymbol{\eta}$

*a. Single species of gas.*

$$\eta = 0.499\rho v_a \lambda \qquad (\text{dyn s cm}^{-2} \text{ or poise}) \tag{3.27}$$

$$= 0.499\, nmv_a\lambda \tag{3.27a}$$

$$= (0.998/\pi\delta_m^2)(mkT/\pi)^{1/2} \tag{3.27b}$$

$$= 2.107 \times 10^{-9}(mT)^{1/2}/\delta_m^2 \tag{3.27c}$$

when $k = 1.381 \times 10^{-16}$ ergs molec.$^{-1}$ K$^{-1}$ in Eq. 3.27b, with $m$ (g), $T$ (K), and $\delta_m$ (cm).

Equation 3.27 is inferred with the assumption that the molecules are smooth, rigid spheres surrounded by fields of attractive force (Dushman, 1966, p. 29).

The units of viscosity in different system of units are

$$1 \text{ poise} = 1 \text{ dyn s cm}^{-2} = 1 \text{ g s}^{-1} \text{ cm}^{-1} \qquad \text{(CGS)}$$
$$1 \text{ Pa s} = 1 \text{ N s m}^{-2} = 1 \text{ kg s}^{-1} \text{ m}^{-1} = 10 \text{ poise} \qquad \text{(SI)}$$

Since at high temperatures $\lambda$ is a function of temperature (see Eq. 3.23h) viscosity becomes also dependent of temperature. Thus

$$\eta_T = (0.998/\pi\delta_m^2)(mkT)(1 + C_s/T) \text{ (poise)} \qquad (3.27d)$$

where $C_s$ is Sutherland's constant (dimensionless). For the values of Sutherland's constant, see Table 3.11.

*The viscosity of gas increases with temperature.*

*b. Mixture of gases.*

$$\eta_{12} = 0.499(n_1 m_1 v_{a1}\lambda_1 + n_2 m_2 v_{a2}\lambda_2) \qquad \text{(poise)} \qquad (3.27e)$$

If the apparent diameters of molecules are $\delta_{m1}$ and $\delta_{m2}$, then

$$\eta_1/\eta_2 = (m_1/m_2)^{1/2} \qquad (3.27f)$$

The values of $\eta$ for various gases and water vapor at different temperatures are listed in Table 3.12.

### 3.11.3 The Collision Frequency $\varepsilon$ per Molecule per Unit Time

$$\varepsilon = v_a/\lambda \qquad (\text{s}^{-1}) \qquad (3.28)$$

$$\varepsilon = 1.273\ p/\eta \qquad (\text{s}^{-1}) \qquad (3.28a)$$

where $p$ (dyn cm$^{-2}$) and $\eta$ (dyn s cm$^{-2}$).

The expression of $\varepsilon$ as a function of pressure and viscosity in different systems of units, is

$$\varepsilon = 1.29 \times 10^6\ p/\eta \qquad (\text{s}^{-1}) \qquad (3.28b)$$

with $p$ (Pa) and $\eta$ (Pa s),

$$\varepsilon = 1273\ p/\eta \qquad (\text{s}^{-1}) \qquad (3.28c)$$

with $p$ (mbar) and $\eta$ (poise), and

$$\varepsilon = 1697\ p/\eta \qquad (\text{s}^{-1}) \qquad (3.28d)$$

with $p$ (Torr) and $\eta$ (poise).

**Table 3.12**

Coefficients of Viscosity $\eta$ and Collision Frequency per Molecule $\varepsilon$, per Unit Time, at Atmospheric Pressure and Various Temperatures[a]

| Gas | $\eta$ ($\mu$poise) $T = 0°C$ | $T = 15°C$ | $T = 23°C$ | $T = 23°C$ $10^{-9}\varepsilon$ ($s^{-1}$) |
|---|---|---|---|---|
| *Monoatomic* | | | | |
| He | 185.4 | 190.4 | 192.4 | 6.70 |
| Ne | 317.3 | 325.8 | 330.7 | 3.91 |
| Ar | 208.4 | 214.1 | 217.0 | 5.94 |
| Kr | 225.3 | 231.3 | 234.5 | 5.50 |
| Xe | 217.4 | 223.6 | 226.3 | 5.70 |
| *Diatomic* | | | | |
| $H_2$ | 88.6 | 91.0 | 92.3 | 13.97 |
| $N_2$ | 166.3 | 170.8 | 173.2 | 7.45 |
| CO | 165.4 | 169.9 | 172.2 | 7.49 |
| $O_2$ | 190.7 | 195.8 | 198.7 | 6.49 |
| Air | 170.7 | 175.3 | 177.7 | 7.26 |
| *Polyatomic* | | | | |
| $CO_2$ | 137.6 | 141.3 | 143.3 | 9.00 |
| $H_2O$[b] vapor | 87.3 | 93.1 | 96.9 | — |

[a] $\eta$ calculated from Eq. 3.27c; $\varepsilon$ calculated from Eq. 3.28c.
[b] Interpolated from the values given in "Viscosity of Gases" table in *Handbook of Chemistry and Physics* (CRC, 1986–1987, p. F-44).

Table 3.12 lists the values of the collision frequency $\varepsilon$ per molecule per unit time for various gases.

### 3.11.4 The Coefficient of Self-Diffusion $D_{1,1}$

The diffusion of molecules of a single gas within the gas mass can be calculated in terms of the coefficient of self-diffusion as

$$D_{1,1} = v_a \lambda / 3 \qquad (\text{e.g., cm}^2\ \text{s}^{-1}) \tag{3.29}$$

$$= 1.342\ \eta / \rho \tag{3.29a}$$

The value 1.342 in Eq. 3.29a results from introducing different distribution factors (Dushman, 1966, p. 66).

*For constant temperature, the coefficient of self-diffusion varies directly with the mean free path $\lambda$ of gas molecules or inversely as the density $\rho$ of gas.*

**Table 3.13**
Observed Coefficients of Interdiffusion $D_{1,2}$ for Several Pairs of Gases at STP and Average Molecular Diameters $\delta^{a}_{m1,2}$

| Gas | $D_{1,2}$ observed ($cm^2\ s^{-1}$) | $10^8\delta_{m1,2}$ (cm) |
|---|---|---|
| $H_2$–air | 0.661 | 3.23 |
| $H_2$–$O_2$ | 0.679 | 3.18 |
| $O_2$–air | 0.1775 | 3.69 |
| $O_2$–$N_2$ | 0.174 | 3.74 |
| CO–$H_2$ | 0.642 | 3.28 |
| CO–$O_2$ | 0.183 | 3.65 |
| $CO_2$–$H_2$ | 0.538 | 3.56 |
| $CO_2$–air | 0.138 | 4.03 |
| $CO_2$–CO | 0.136 | 4.09 |
| $N_2O$–$H_2$ | 0.535 | 3.57 |
| $N_2O$–$CO_2$ | 0.0983 | 4.53 |

[a]Adapted from Dushman (1966, p. 68).

### 3.11.5 The Coefficient of Interdiffusion $D_{1,2}$

*a. Gases consisting of different molecules 1, 2.*

$$D_{1,2} = \left(v_{a1}^2 + v_{a2}^2\right)^{1/2} / \left(3\pi n_{1,2}\delta_{m1,2}^2\right) \quad \text{(e.g., cm}^2\ \text{s}^{-1}) \quad (3.30)$$

where $n_{1,2} = n_1 + n_2$ and $\delta_{m1,2} = (\delta_{m1} + \delta_{m2})/2$.

When the number density of molecules of one of the gases is very small (traces) such that $n_2 \ll n_1$, then

$$D_{1,2} = [2/(3\delta_{m1,2}p)]\left[k^3T^3/(\pi^3 m)\right]^{1/2} \quad (3.30a)$$

The principle underlying the functioning of diffusion pumps is the interdiffusion of a gas having a low number density of molecules into another gas that has a very high number density of molecules.

Table 3.13 lists the observed coefficients of interdiffusion for several pairs of gases at STP and the molecular diameter calculated from Eq. 3.30.

*b. Gases consisting of identical molecules.*

$$D_{1,1} = (2/3)v_a/\left(2^{1/2}\pi n_{1,2}\delta_m^2\right) \quad \text{(e.g., cm}^2\ \text{s}^{-1}) \quad (3.30b)$$

where $n_{1,2} = n_1 + n_2$.

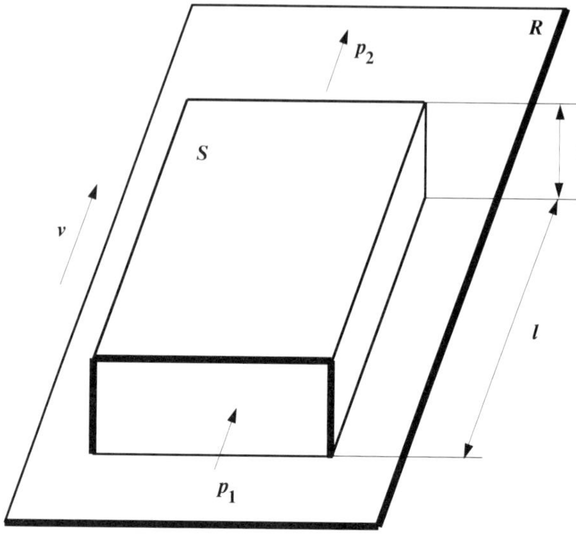


**Figure 3.4** Schematic of the principle of molecular drag pumping ($S$ = stationary groove of length $l$ and depth $h$; $R$ = planar surface moving at speed $v$ with respect to the groove $S$; $p_1$, $p_2$ = pressures at the inlet and outlet from the groove).

## *3.12 Transport Phenomena in the Molecular State*

At low pressures, the mean free path $\lambda$ of molecules becomes very large compared to the dimensions of the enclosure that contains the gas. The transport of energy from wall to wall does not include molecular collisions and is, therefore, no longer a function of viscosity.

### 3.12.1 Molecular Drag

For the steady state of gas (zero flow) between a stationary groove and a plane surface $R$ (Fig. 3.4) moving at speed $v$ with respect to the groove, the pressure $p_1$ at the inlet of the groove increases to $p_2$ in the direction of the plane surface movement, when $\lambda \gg l$ (where $l$ is the length of the groove). The ratio of pressures $p_2$ at the outlet and $p_1$ at the inlet from the groove is

$$p_2/p_1 = \exp(M/2\pi RT)^{1/2}(vl/h) \tag{3.31}$$

with M (g mol$^{-1}$), $R$ (erg mol$^{-1}$ K$^{-1}$), $T$ (K), $v$ (cm s$^{-1}$), $l$ (cm), and $h$ (cm).

The principle of molecular drag underlies the function of molecular gauges (see also Berman, 1985, pp. 134–136)

## 3.13 *The Ratio ν of Specific Heat Capacities*

If $c_v$ and $c_p$ are the heat capacities at constant volume and constant pressure, respectively, of an ideal gas, then

$$\nu = c_p/c_v \qquad \text{(dimensionless)} \tag{3.32}$$

$$= 1.67 \text{ for monoatomic gases} \tag{3.32a}$$

$$= 1.40 \text{ for diatomic gases} \tag{3.32b}$$

$$= 1.33 \text{ for polyatomic gases} \tag{3.32c}$$

The unit of $c_p$ or $c_v$ in the SI system is J $kg^{-1}$ $K^{-1}$.

## 3.14 *Heat Conductivity of Rarefied Gases*

### 3.14.1 Heat Conductivity in the Viscous Range

For an ideal gas, the coefficient of heat conductivity $H_v$ is

$$H_v = (1/3)(\rho v_a c_v \lambda) \qquad (\text{e.g., J m}^{-1}\text{ s}^{-1}\text{ K}^{-1}) \tag{3.33}$$

$$= (1/4)(9\nu - 5)(\eta c_v) \tag{3.33a}$$

with $\rho$ (kg $dm^{-3}$), $v_a$ (m $s^{-1}$), $c_v$ (J $kg^{-1}$ $K^{-1}$), $\lambda$ (m), and $\eta$ (Pa s or N s $m^{-2}$)

*The heat conductivity of gases is independent of pressure in the viscous range of flow.*

Table 3.14 lists the coefficient of heat conductivity $H_v$ for some gases and mercury vapor as well as the experimental ratio $\nu$ for these gases.

### 3.14.2 Heat Conductivity in the Molecular Range

The process of heat transfer in the molecular range of flow (where the mean free path of molecules $\lambda$ is much larger than the vessel that accommodates the gas) is characterized by the accommodation coefficient $\alpha$.

$$\alpha = (T_r - T_g)/(T_s - T_g) \qquad \text{(dimensionless)} \tag{3.34}$$

where $T_g$ (K) is the temperature of a group of gas molecules (atoms) that strike a surface of temperature $T_s$ (K), ($T_s > T_g$) and where $T_r$ (K) is the temperature of the gas molecules reflected from the surface.

**Table 3.14**
The Coefficient of Heat Conductivity $H_v$ and the Ratio $\nu = c_p/c_v$ for Some Gases and Hg Vapor at $T = 0°C$[a]

| | Monoatomic | | | | |
|---|---|---|---|---|---|
| $H_v$[b] (cal cm$^{-1}$ s$^{-1}$ K$^{-1}$) | He ($3.43 \times 10^{-4}$) | Ar ($3.90 \times 10^{-5}$) | Kr ($2.10 \times 10^{-5}$) | Ne ($1.09 \times 10^{-4}$) | Hg ($1.20 \times 10^{-5}$) |
| $\nu$ experimental | 1.67 | 1.67 | 1.69 | 1.67 | 1.67 |
| $\nu$ ideal | | | 1.67 | | |
| | Diatomic | | | | |
| $H_v$ (cal cm$^{-1}$ s$^{-1}$ K$^{-1}$) | $H_2$ ($4.19 \times 10^{-4}$) | $N_2$ ($5.70 \times 10^{-5}$) | $O_2$ ($5.80 \times 10^{-5}$) | CO ($5.30 \times 10^{-5}$) | Air ($5.80 \times 10^{-5}$) |
| $\nu$ experimental | 1.41 | 1.40 | 1.40 | 1.404 | 1.403 |
| $\nu$ ideal | | | 1.40 | | |
| | Polyatomic | | | | |
| $H_v$ (cal cm$^{-1}$ s$^{-1}$ K$^{-1}$) | $CO_2$ ($3.40 \times 10^{-5}$) | | | | |
| $\nu$ experimental | 1.30 | | | | |
| $\nu$ ideal | 1.33 | | | | |

[a]Adapted from Roth (1982, p. 57).
[b]*Note*: To convert $H_v$ (cal cm$^{-1}$ s$^{-1}$ K$^{-1}$) to $H_v$ (J m$^{-1}$ s$^{-1}$ K$^{-1}$), multiply by 4.187 J cal$^{-1}$.

The accommodation coefficient $\alpha$ can also be expressed in terms of the total mean energy of the gas molecules (atoms). Thus

$$\alpha = (E_r - E_g)/(E_s - E_g) \qquad \text{(dimensionless)} \qquad (3.34a)$$

The values of the accommodation coefficient $\alpha$ of gases on several surfaces are listed in Table 3.15.

## 3.15 *Energy Transport in Rarefied Gases*

### 3.15.1 Average Energy Transferred per Molecule $\dot{E}$

$$\dot{E} = (2/3)m\bar{v}_r^2 \qquad (\text{J molec.}^{-1}) \qquad (3.35)$$

[*Note*: The average energy *transferred* per molecule $\dot{E}$ (Eq. 3.35) differs from the average translational energy per molecule $\overline{E}$ in random motion (Eq. 3.9a) by a factor of 4/3.]

**Table 3.15**
Values of the Accommodation Coefficient $\alpha$

| Gas \ Metal | Pt: Slightly covered with black | Pt: Heavily covered with black | W | Fe | Ni |
|---|---|---|---|---|---|
| *Monoatomic* | | | | | |
| He | — | — | 0.0164[a] | — | — |
| Ne | — | — | — | 0.1[c] | 0.82[b] |
| | 0.89[d] | — | 0.85[d] | — | 0.93[b] |
| Ar | — | — | 0.89[e] | — | — |
| Kr | 0.69[f] | — | — | — | — |
| Hg | — | — | 0.95[d] | — | — |
| *Diatomic* | | | | | |
| $H_2$ | 0.28[b] | 0.71[d] | 0.36[b] | — | 0.29[b] |
| | 0.36[d] | — | 0.2[d] | — | — |
| $N_2$ | 0.89[d] | — | 0.57[d] | — | 0.82[b] |
| | 0.81[b] | — | 0.87[b] | — | — |
| $O_2$ | 0.85[b] | 0.95[d] | 0.90[b] | — | 0.86[b] |
| *Polyatomic* | | | | | |
| Air | 0.90[d] | — | — | — | — |
| $CO_2$ | — | 0.97[d] | — | — | — |

[a]After Thomas and Schofield (1955).
[b]After Armdur and Guildner (1957).
[c]After Eggelton and Tompkins (1952).
[d]After Dushman (1966).
[e]After De Porter and Searcy (1963).
[f]After Thomas and Brown (1950).

### 3.15.2 Energy Transferred by Molecules $E_v$ in the Viscous Range

*a. Between two plates at temperatures $T_1$ and $T_2 (T_1 > T_2)$ a distance d apart.* The rate of energy transferred is

$$E_v = (1/3)nmv_a c_v(T_1 - T_2)(\lambda/d) \qquad (\text{e.g., J m}^{-2}\text{ s}^{-1}) \quad (3.36)$$

$$= (1/3)(\rho v_a c_v)(\lambda/d)(T_1 - T_2) \qquad (\text{e.g., W m}^{-2}) \quad (3.36a)$$

with $\rho$ (kg m$^{-3}$), $v_a$(m s$^{-1}$), $c_v$ (J kg$^{-1}$ K$^{-1}$), $\lambda$ (m), $d$ (m), and $T_1$, $T_2$ (K).

To convert $E_v$ (Jm$^{-2}$ s$^{-1}$) to $E_v$ (cal m$^{-2}$ s$^{-1}$), multiply by 0.2338 (cal J$^{-1}$).

*b. From a hot wire of radius $r_w$, length $l_w$, and temperature $T_w$ suspended along the axis of a cylinder of radius $r_{cyl}$ and temperature $T_{cyl}$.*

(i) Per unit time:

$$E_{vt} = 2\pi l_w H_v(T_w - T_{cyl})/\ln(r_{cyl}/r_w) \qquad (\text{J s}^{-1}\text{ or W}) \quad (3.37)$$

(ii) Per unit area of the wire:

$$E_{va} = H_v(T_w - T_{cyl})/r_w \ln(r_{cyl}/r_w) \qquad (\text{J m}^{-2}\ \text{s}^{-1}) \quad (3.38)$$

(iii) Per unit length of the wire:

$$E_v = 2\pi H_v(T_w - T_{cyl})/\ln(r_{cyl}/r_w) \qquad (\text{J m}^{-1}\ \text{s}^{-1}\ \text{or W m}^{-1}) \quad (3.39)$$

with $l_w$ (m) and $H_v$(J m$^{-1}$ s$^{-1}$ K$^{-1}$) calculated from Eq. 3.33 as the average heat conductivity over the range $T_w \cdots T_{cyl}$ (K) and $r_{cyl}$, $r_w$ (m).

Equations 3.37–3.39 hold true for $100 < (r_{cyl}/r_w) < 300$.

### 3.15.3 Energy Transferred by Molecules $E_m$ in the Molecular Range

*a. From a hot surface (at temperature $T_s$) to a cold surface (at temperature $T_g$) per square centimeter of the hot surface, per second.* Monoatomic gases that acquire only translational energy (subscript 1) are

$$E_{m1} = (1/4)2\ knv_{ag}(T_r - T_g) \qquad (\text{W cm}^{-2}) \qquad (3.40)$$

$$= (\alpha/2)pv_{ag}(T_s - T_g)/T_g \qquad (3.40a)$$

where $v_{ag}$ (cm s$^{-1}$) is calculated from Eq. 3.2 for $T_g = T$, with $p$ (dyn cm$^{-2}$), $T_r$, $T_g$, $T_s$ (K) (see Eq. 3.34), and where $\alpha$ (dimensionless) is the accommodation coefficient.

Diatomic and polyatomic gases that acquire, in addition to translational energy (subscript 1) rotational (subscript 2) and vibrational (subscript 3) energy, respectively, are

$$E_{m1,2,3} = (\alpha_s/8)\Gamma pv_{ag}(T_s - T_g)/T_g \qquad (\text{W cm}^{-2}) \qquad (3.41)$$

$$= \alpha_s\Lambda_0 p(273.2/T_g)^{1/2}(T_s - T_g) \qquad (3.41a)$$

where $\alpha_s = (\alpha_1\alpha_2)/(\alpha_1 + \alpha_2 - \alpha_1\alpha_2)$ (dimensionless) and $\alpha_1, \alpha_2$ (dimensionless) are the accommodation coefficients for the hot and cold surfaces, respectively, and

$$\Gamma = (\nu + 1)/(\nu - 1) \qquad (\text{dimensionless})$$

$$\Lambda_0 = 1.468 \times 10^{-5}\ M^{-1/2} \qquad (\text{W cm}^{-2}\ \text{K}^{-1}\ \mu\text{m Hg}^{-1})$$

with $p$ (dyn cm$^{-2}$), $v_{ag}$ (cm s$^{-1}$) calculated from Eq. 3.1 for $T_g = T$, $T_s$, $T_g$ (K), and where $\Lambda_0$ is the free molecular conductivity at 0°C. The values of $\Lambda_0$ for some gases, water, and mercury vapor are listed in Table 3.16.

*In the molecular range, the rate of energy transferred by molecules between two surfaces is proportional to the pressure and the temperature difference between the surfaces.*

**Table 3.16**
Values of the Molecular Free Conductivity $\Lambda_0$ for Some Gases and Hg Vapor

| Gas | M ($g\ mol^{-1}$) | $\nu$ | $10^6\ \Lambda_0$ ($W\ cm^{-2}\ K^{-1}\ \mu m\ Hg^{-1}$)[b] |
|---|---|---|---|
| $H_2$ | 2.016 | 1.41 | 60.72 |
| He | 4.003 | 1.67 | 29.35 |
| $H_2O$ | 18.016 | 1.30 | 26.49 |
| Ne | 20.18 | 1.67 | 13.07 |
| $N_2$ | 28.02 | 1.40 | 16.63 |
| $O_2$ | 32.00 | 1.40 | 15.57 |
| Ar | 39.94 | 1.67 | 9.29 |
| $CO_2$ | 44.01 | 1.30 | 16.96 |
| Hg | 200.6 | 1.67 | 4.15 |

[a]Adapted from Dushman (1966, p. 48).
[b]*Note:* 1 mm Hg = 1 Torr.

*b. From a hot wire of radius $r_2$ and temperature $T_s$ to a cylinder of radius $r_1$ ($r_1 > r_2$) and temperature $T_g$ ($T_s > T_g$), when the wire is suspended along the axis of the cylinder.*

$$E_{m,\,cyl} = \alpha_r \Lambda_0 p(273.2)/T_g)^{1/2}(T_s - T_g) \qquad (W\ cm^{-2}) \qquad (3.42)$$

where $\alpha_r = \alpha/[1 - (1 - \alpha)(r_2/r_1)]$ (dimensionless) with $\Lambda_0$($W\ cm^{-2}\ K^{-1}$ $\mu$bar), $p$ ($dyn\ cm^{-2}$), and $T_g$, $T_s$ (K).

## 3.16 *Thermomolecular Flow*

### 3.16.1 Thermal Transpiration

If two chambers (1 and 2) separated by a porous plug or a narrow pipe contain gas at temperatures $T_1$ and $T_2$, respectively, then

$$p_1 T_2^{1/2} = p_2 T_1^{1/2} \qquad (3.43)$$

Gas will flow between the two chambers until an equilibrium state is established.

A more accurate approach (Takaishi and Sensui, 1963) gives an agreement within 0.5% between the calculated curve from Eq. 3.43 and the experimental

**Table 3.17**

Values[a] of Coefficients $A^*$, $B^*$, $C^*$ in Equations 3.43a and 3.43b for Some Gases[b]

| Gas | $10^{-6}A^*$ ($K^2\ Pa^{-2}\ m^{-2}$) | $10^{-3}B^*$ ($K\ Pa^{-1}\ m^{-1}$) | $C^*$ [$K^{1/2}\ Pa^{-(1/2)}\ m^{-(1/2)}$] | $10^{-5}A^*$ ($K^2\ mm\ Hg^{-2}\ mm^{-2}$) |
|---|---|---|---|---|
| $H_2$ | 9.68 | 6.001 | 29.03 | 1.24 |
| Ne | 14.91 | 1.41 | 82.16 | 2.65 |
| Ar | 60.75 | 6.06 | 42.72 | 10.80 |
| Kr | 81.56 | 11.25 | 37.52 | 14.50 |
| $CH_4$ | 81.56 | 11.25 | 35.60 | 14.50 |
| Xe | 196.88 | 31.05 | 27.39 | 35.00 |
| He | 8.44 | 0.86 | 52.04 | 1.50 |
| $N_2$ | 67.50 | 7.50 | 38.34 | 12.00 |

| Gas | $10^{-2}B^*$ ($K\ mm\ Hg^{-1}\ mm^{-1}$) | $C^*$ [$K^{1/2}\ mm\ Hg^{-(1/2)}\ mm^{-(1/2)}$] | Temperature range (K) |
|---|---|---|---|
| $H_2$ | 8.00 | 10.6 | 14–673 |
| Ne | 1.88 | 30.0 | 20.4–673 |
| Ar | 8.08 | 15.6 | 77–673 |
| Kr | 15.00 | 13.7 | 77–673 |
| $CH_4$ | 15.00 | 13.0 | 473–673 |
| Xe | 41.40 | 10.0 | 77–90 |
| He | 1.15 | 19.0 | 42–90 |
| $N_2$ | 10.00 | 14.0 | 77–195 |

[a]All data have been obtained in a glass apparatus.
[b]Adapted from Takaishi and Sensui (1963, p. 2511).

results. Thus, for $T_2 < T_1$

$$p_2/p_1 = \left[A^*(X/\bar{T})^2 + B^*(X/\bar{T}) + C^*(X/\bar{T})^{1/2} + (T_2/T_1)^{1/2}\right] \Big/ \left[A^*(X/\bar{T})^2 + B^*(X/\bar{T}) + C^*(X/\bar{T})^{1/2} + 1\right] \tag{3.43a}$$

and for $T_2 > T_1$

$$p_1/p_2 = \left[A^*(X/\bar{T})^2 + B^*(X/\bar{T}) + C^*(X/\bar{T})^{1/2} + (T_1/T_2)^{1/2}\right] \Big/ \left[\left(A^*(X/\bar{T})^2 + B^*(X/\bar{T}) + C^*(X/\bar{T})^{1/2} + 1\right]\right. \tag{3.43b}$$

where the subscript 2 refers to the quantities at room temperature; $A^*$, $B^*$, $C^*$ are constants specific to the gas concerned, independent of temperature (their values are listed in Table 3.17); and where

$\overline{T} = (T_1 + T_2)/2$ (K)
$X = p_2 d$ (Pa m)
$d$ (m)= inner diameter of the tube that connects the manometer that measures $p_2$ to the vacuum system

For gases other than those listed in Table 3.17, the value of the constants $A^*$, $B^*$, $C^*$ can be calculated from the following equations:

$$A^* = 1.4 \times 10^4 \exp(0.507\, \delta_m) \qquad (\mathrm{K^2\ mm\ Hg^{-1}\ mm^{-2}}) \quad (3.43c)$$

$$B^* = 5.6 \exp(0.607\, \delta_m) \qquad (\mathrm{K\ mm\ Hg^{-1}\ mm^{-1}}) \quad (3.43d)$$

$$C^* = (110/\delta_m) - 14 \qquad [\mathrm{K^{1/2}\ mm\ Hg^{-(1/2)}\ mm^{-(1/2)}}] \quad (3.43e)$$

The values of $\delta_m$ calculated from viscosity data at 25°C in Eq. 3.27b must be multiplied with $10^{-8}$ (cm) before substituting them in Eqs. 3.43c–3.43e. *These equations lack theoretical basis and therefore have to be used carefully.*

For thermal transpiration the corrections of the pressures measured with capacitance manometers are given in Poulter et al. (1983), Buckman et al. (1984), and Jitschin and Röhl (1987).

## *Exercises*

**Exercise 3.1**

Estimate the order of magnitude of the pressure for which the van der Waals equation of state describes satisfactorily the behavior of gas at $T = 23°\mathrm{C}$.

**Solution 3.1**

Assume that the distance $L$ between molecules is equal to their apparent diameter $\delta_m$, and that the mean diameter of a molecule is $2.0 \times 10^{-10}$ m. Therefore, $L = 2.0 \times 10^{-10}$ m and the resulting density of molecules (see Exercise 1.20) is

$$n = L^{-3} = (2.0 \times 10^{-10}\ \mathrm{m})^{-3} = 1.25 \times 10^{29}\ \mathrm{molec.\ m^{-3}}$$

Use Eq. 1.15b

$$p = nkT$$

and substitute

$k = 1.381 \times 10^{-23}$ J molec.$^{-1}$ K$^{-1}$
$T = 273 + 23 = 296$ K

Thus

$$p = 1.25 \times 10^{29} \text{ molec. m}^{-3} \times 1.381 \times 10^{-23} \text{ J molec.}^{-1} \text{ K}^{-1} \times 296 \text{ K}$$

$$= 5.1 \times 10^{8} \text{ N m}^{-2} = 5.1 \times 10^{8} \text{ Pa} = 5.1 \text{ Pa}/1.01 \times 10^{5} \text{ Pa atm}^{-1}$$

$$= 5050 \text{ atm}$$

**Exercise 3.2**

Calculate the number of molecules per cubic centimeter (molec. $cm^{-3}$) at $T = 23°C$ and at pressure $p = 1.0 \times 10^{-12}$ mbar.

**Solution 3.2**

Use Eq. 1.8d

$$n = 7.243 \times 10^{21}\, p/T \qquad (\text{molec. dm}^{-3})$$

and substitute

$p = 1.0 \times 10^{-12}$ mbar
$T = 273 + 23 = 296$ K

Thus

$$n = 7.243 \times 10^{21} \times 1.0 \times 10^{-12}/296 = 2.4 \times 10^{7} \text{ molec. dm}^{-3}$$

$$= 2.4 \times 10^{4} \text{ molec. cm}^{-3}$$

**Exercise 3.3**

Calculate (a) the arithmetic average velocity $v_a$ and (b) the root-mean-square velocity $v_r$ of nitrogen molecules in a trap cooled at $T_1 = -196°C$ (liquid nitrogen), $T_2 = -78°C$ (dry ice + acetone mixture), and $T_3 = 23°C$ in the SI system of units.

**Solution 3.3**

Use Eqs. 3.2c and 3.4e

$$v_a = 4.602(T/M)^{1/2} \qquad (\text{m s}^{-1})$$

$$v_r = 4.994(T/M)^{1/2} \qquad (\text{m s}^{-1})$$

and substitute

$M_{N_2} = 28 \times 10^{-3}$ kg mol$^{-1}$ (see Table 1.2)
$T_1 = 273 - 196 = 77$ K
$T_2 = 273 - 78 = 195$ K
$T_3 = 273 + 23 = 296$ K

Thus

(a) $$v_{a(77\ K)} = 4.602(77/28 \times 10^{-3})^{1/2} = 241\ m\ s^{-1}$$

$$v_{a(195\ K)} = 4.602(195/28 \times 10^{-3})^{1/2} = 384\ m\ s^{-1}$$

$$v_{a(296\ K)} = 4.602(296/28 \times 10^{-3})^{1/2} = 473\ m\ s^{-1}$$

(b) $$v_{r(77\ K)} = 4.994(77/28 \times 10^{-3})^{1/2} = 262\ m\ s^{-1}$$

$$v_{r(195\ K)} = 4.994(195/28 \times 10^{-3})^{1/2} = 417\ m\ s^{-1}$$

$$v_{r(296\ K)} = 4.994(296/28 \times 10^{-3})^{1/2} = 513\ m\ s^{-1}$$

**Exercise 3.4**

Tubulated ionization gauge heads are used to measure "pressure" in ultra-high-vacuum systems. Calculate the "pressure" measured by such a gauge head when (a) the temperatures of both the gauge head and the system are reduced to the same extent and (b) the temperature of the gauge head is substantially reduced with respect to that of the system.

**Solution 3.4**

A tubulated gauge head consists of a system of electrodes enclosed in a shell, connected at the location where pressure is to be measured. The residual gas existing in the shell is ionized to a certain amount, and the ions produced are collected and measured as a current in an electric circuit. This current provides an indication of the molecular density of the residual gas. Thus, *ionization gauges measure actually gas density*.

(a) If the temperatures of both the ionization gauge head and vacuum system are $T_1$ and are then reduced to $T_2$ such that $T_2 < T_1$, the root-mean-square velocity of molecules (see Eq. 3.4a) would be $v_{r \text{ at } T_2} < v_{r \text{ at } T_1}$. The net rate of transfer of momentum by gas molecules will decrease in both the system and the gauge head. The number density of molecules $n$ would change nearly at the same rate in both the system and the gauge head. Thus $p_{2 \text{ at } T_2} < p_{1 \text{ at } T_1}$.

(b) If the temperature of the gauge head is reduced to $T_2$ and the temperature to the system $T_1$ is higher, the number density of molecules $n_2$ in the gauge head would be higher than the number density of molecules $n_1$ in the system, since $n$ is inversely proportional to the temperature $T$ (see Eq. 1.8b). Thus the gauge head will measure a "pressure" $p_2$ higher than the pressure $p_1$ in the system. Thus $p_{2 \text{ at } T_2} > p_{1 \text{ at } T_1}$.

**Exercise 3.5**

Calculate the total translational energy $E$ of molecules in random motion in a mole of ideal gas at $T = 23°C$ in joules, calories, and electron volts.

**Solution 3.5**

From Eq. 3.11 written for 1 mol

$$E = (3/2)RT \qquad \text{J mol}^{-1}$$

On substituting

$R = 8.314 \text{ J mol}^{-1} \text{ K}^{-1}$ (see Table 1.3)
$T = 273 + 23 = 296 \text{ K}$
$1 \text{ J} = 0.2338$ cal (see Appendix, Table A.16)
$= 6.242 \times 10^{18}$ eV (see Appendix, Table A.16)

the value of $E$ results:

$$\begin{aligned} E &= (3/2) \times 8.314 \text{ J mol}^{-1} \text{ K}^{-1} \times 296 \text{ K} = 3691 \text{ J mol}^{-1} \\ &= 3691 \text{ J mol}^{-1} \times 0.2338 \text{ cal J}^{-1} = 863 \text{ cal mol}^{-1} \\ &= 3691 \text{ J mol}^{-1} \times 6.242 \times 10^{18} \text{ eV J}^{-1} = 2.3 \times 10^{22} \text{ eV mol}^{-1} \end{aligned}$$

**Exercise 3.6**

Calculate for 1 mol of ideal gas (a) the average translational energy $\bar{E}$ per molecule in random motion at $T = -196°C$ (temperature of boiling liquid nitrogen) as well as at $T = 23°C$ and (b) the total translational energy $E$ of molecules in random motion at the temperatures specified at (a).

**Solution 3.6**

(a) Use Eq. 3.9a

$$\bar{E} = (3/2)(kT)$$

and substitute

$k = 1.381 \times 10^{-23} \text{ J molec.}^{-1} \text{ K}^{-1}$ (see Table 1.3)
$T = 273 - 196 = 77 \text{ K}$
$T = 273 + 23 = 296 \text{ K}$

Thus

$$\begin{aligned} \bar{E}_{77\text{ K}} &= (3/2)(1.381 \times 10^{-23} \text{ J molec.}^{-1} \text{ K}^{-1} \times 77 \text{ K}) \\ &= 1.595 \times 10^{-21} \text{ J molec.} \\ \bar{E}_{296\text{ K}} &= (3/2)(1.381 \times 10^{-23} \text{ J molec. K}^{-1} \times 296 \text{ K}) \\ &= 6.132 \times 10^{-21} \text{ J molec.}^{-1} \end{aligned}$$

(b) The total translational energy $E$ of molecules in random motion can be expressed as

$$E = \bar{E}N_A$$

Substitute the values of $\bar{E}$ calculated at (a) and

$$N_A = 6.022 \times 10^{23} \text{ molec. mol}^{-1} \qquad \text{(see Section 1.9)}$$

Thus

$$E_{77\text{ K}} = 1.595 \times 10^{-21} \text{ J molec.} \times 6.022 \times 10^{23} \text{ molec. mol}^{-1}$$

$$= 960 \text{ J molec.}^{-1}$$

$$E_{296\text{ K}} = 6.132 \times 10^{-21} \text{ J molec.} \times 6.022 \times 10^{23} \text{ molec. mol}^{-1}$$

$$= 3693 \text{ J molec.}^{-1}$$

**Exercise 3.7**

Compare the *average energy transferred per molecule* $\dot{E}$ and the *average translational energy* $\bar{E}$ *per molecule* in random motion.

**Solution 3.7**

The average translational energy $\bar{E}$ per molecule in random motion is (Eq. 3.9a)

$$\bar{E} = (3/2)(kT) \qquad \text{J molec.}^{-1}$$

The average energy transferred per molecule $\dot{E}$ (Eq. 3.35) is

$$\dot{E} = (2/3)mv_r^2 \qquad \text{J molec.}^{-l}$$

Substitute the value of $v_r$ (the mean-square velocity) by

$$v_r = (3\ kT/m) \qquad \text{(Eq. 3.3a)}$$

Thus

$$\dot{E} = (2/3)m(3\ kT/m) = 2\ kT$$

$$\dot{E}/\bar{E} = 2\ kT/[(3/2)(kT)] = 4/3$$

**Exercise 3.8**

Use the coefficient of viscosity $\eta$ at $T = 0°\text{C}$ to calculate the diameter of a molecule of neon.

**Solution 3.8**

From Eq. 3.20

$$\delta_m = \left(2.107 \times 10^{-9}\,(mT)^{1/2}/\eta\right)^{1/2} \qquad \text{cm}$$

and with

$$m = 3.51 \times 10^{-26} \text{ kg molec.}^{-1} = 3.51 \times 10^{-23} \text{ g molec.}^{-1} \text{ (see Table 1.2)}$$
$$T = 273 + 0 = 273 \text{ K}$$
$$\eta_{0°C} = 317.3 \times 10^{-6} \text{ poise (see Table 3.12)}$$
$$\delta_m = 2.107 \times 10^{-9} \times (3.51 \times 10^{-23} \times 273)^{1/2}/317.3 \times 10^{-6})^{1/2}$$
$$= 2.55 \times 10^{-8} \text{ cm}$$

**Exercise 3.9**

A cylindrical tube of 1 mm diameter contains argon at $T = 15°C$. Determine the collision frequency $\varepsilon$ per molecule per unit time and the rate $\phi$ at which molecules at steady state strike a unit area of the tube per unit time, when the value of the Knudsen number is (a) $K' = 10^4$ and (b) $K' = 5 \times 10^4$.

**Solution 3.9**

To calculate $\varepsilon$ and $\phi$ it is necessary to first determine the arithmetic average velocity $v_a$, the mean free path $\lambda$, and the number density $n$ of molecules. Hence, use

$$v_a = (8RT/\pi M)^{1/2} \text{ (Eq. 3.2)}$$
$$\lambda = K'd \text{ (Eq. 3.26)}$$
$$n = 1/(2^{1/2}\pi\lambda\delta_m^2) \text{ (Eq. 3.23)}$$

Substitute

$$R = 8.314 \times 10^7 \text{ ergs mol}^{-1} \text{ K}^{-1} = 8.314 \times 10^7 \text{ g cm}^2 \text{ s}^{-1} \text{ mol}^{-1} \text{ K}^{-1} \text{ (see Table 1.3)}$$
$$T = 273 + 15 = 288 \text{ K}$$
$$M_{Ar} = 39.9 \times 10^{-3} \text{ kg mol}^{-1} \text{ (see Table 1.2)}$$
$$d = 0.1 \text{ cm}$$
$$\delta_{m,\,Ar} = 3.69 \times 10^{-8} \text{ cm (see Table 3.6)}$$

Thus

$$v_a = \left[(8 \times 8.314 \times 10^7 \text{ g cm}^2 \text{ s}^{-2} \text{ mol}^{-1} \text{ K}^{-1})/(3.14 \times 39.9 \text{ g mol}^{-1})\right]$$
$$= 39102 \text{ cm s}^{-1}$$

Case (a): $\lambda = 10^4 \text{ (dimensionless)} \times 0.1 \text{ cm} = 10^3 \text{ cm}$

$$n = 1/[2^{1/2} \times 3.14 \times 10^4 \times (3.69 \times 10^{-8})^2]$$
$$= 1.66 \times 10^{10} \text{ molec. cm}^{-3}$$

Cases (b): $\lambda = 5 \times 10^{-4} \text{ (dimensionless)} \times 0.1 \text{ cm} = 5 \times 10^{-3} \text{ cm}$

$$n = 1/[2^{1/2} \times 3.14 \times 5 \times 10^{-3} \times (3.69 \times 10^{-8})^2]$$
$$= 3.3 \times 10^{16} \text{ molec. cm}^{-3}$$

Use Eqs. 3.13 and 3.28

$$\phi = nv_a/4$$

$$\varepsilon = v_a/\lambda$$

and substitute

(a) $n = 1.66 \times 10^{10}$ molec. cm$^{-3}$
$v_a = 39102$ cm s$^{-1}$

Thus

$\phi = 1.66 \times 10^{10}$ molec. cm$^{-3}$ $\times$ 39102 cm s$^{-1}/4 = 1.62 \times 10^{14}$ molec. cm$^{-2}$ s$^{-1}$

$\varepsilon = 39102$ cm s$^{-1}/10^3$ cm $= 39.1$ s$^{-1}$

(b) $n = 3.3 \times 10^{16}$ molec. cm$^{-3}$
$v_a = 39102$ cm s$^{-1}$

Thus

$$\phi = 3.3 \times 10^{16} \text{ molec. cm}^{-3} \times 39102 \text{ cm s}^{-1}/4$$

$$= 3.2 \times 10^{20} \text{ molec. cm}^{-2} \text{ s}^{-1}$$

$$\varepsilon = 39102 \text{ cm s}^{-1}/5 \times 10^{-3} \text{ cm} = 7.8 \times 10^{6} \text{ s}^{-1}$$

A comparison of the calculated values shows the following:

| Case | Quantities considered | Remarks |
|---|---|---|
| (a) | $\lambda$ ($10^3$ cm) $\gg d$ (0.1 cm)<br>$\varepsilon = 39.1$ s$^{-1}$<br>$\phi = 1.62 \times 10^{14}$ molec. cm$^{-2}$ s$^{-1}$ | The gas flow is in the molecular range (see Table 4.3) |
| (b) | $\lambda$ ($5 \times 10^{-3}$ cm) $< d$ (0.1 cm)<br>$\varepsilon = 7.8 \times 10^{-6}$ s$^{-1}$<br>$\phi = 3.2 \times 10^{20}$ molec. cm$^{-2}$ s$^{-1}$ | The gas flow is in the transition range (see Table 4.3) |

**Exercise 3.10**

Determine the temperature for which the average translational energy per molecule in random motion $\bar{E} = 1$ eV.

**Solution 3.10**

Use Eq. 3.9a written as

$$T = (2/3)(\bar{E}/k) \qquad \text{K}$$

and substitute

1 eV $= 1.602 \times 10^{-19}$ J (see Appendix, Table A.16)
$k = 1.381 \times 10^{-23}$ J molec.$^{-1}$ K$^{-1}$ (see Table 1.3)

Thus

$$T = (2/3)(1.602 \times 10^{-19}\ \text{J molec.}^{-1}/1.381 \times 10^{-23}\ \text{J molec.}^{-1}\ \text{K}^{-1})$$
$$= 7734\ \text{K} = 7461^{\circ}\text{C}$$

**Exercise 3.11**

A spherical chamber of 50 cm diameter is evacuated at a very low pressure, then flushed with nitrogen and again succesively evacuated at 10, $10^{-3}$, $10^{-6}$, $10^{-9}$ mbar. The temperature of the sphere is $T = 23^{\circ}\text{C}$ (constant) in all these situations. Calculate the ratio $r$ of the molecules in a monolayer on the inner surface of the sphere to the molecules in the volume of the sphere.

**Solution 3.11**

$$r = [n_m\ (\text{molec. cm}^{-2})/n(\text{molec. cm}^{-3})] \times A_s/V_s \qquad \text{(dimensionless)}$$

where $n_m$ = the number of molecules adsorbed per square centimeter to form a monolayer (Eq. 3.22f)
$n$ = the number density of molecules (Eq. 1.8d)
$A_s$ = the total inner surface of the sphere
$V_s$ = the volume of the sphere

(i) Use Eqs. 3.22f and 1.8d

$$n_m = 1/(1.035\pi\delta_m^2/4)\ \text{molec. cm}^{-2}$$
$$n = 7.243 \times 10^{21} p/T\ \text{molec. dm}^{-3}$$

and substitute

$$\delta_m = 3.78 \times 10^{-9}\ \text{dm (see Table 3.6)}$$
$$p = 1.0,\ 10^{-3},\ 10^{-6},\ 10^{-9}\ \text{mbar}$$
$$T = 273 + 23 = 296\ \text{K}$$

Thus

$$n_m = 1/\left[1.035 \times 3.14 \times (3.78 \times 10^{-9}\ \text{dm})^2/4\right]$$
$$= 8.16 \times 10^{16}\ \text{molec. dm}^{-2}$$

For $p = 1.0$ mbar

$$n = 7.243 \times 10^{21} \times 1.0/296 = 2.4 \times 10^{19}\ \text{molec. dm}^{-3}$$

For $p = 10^{-3}$ mbar

$$n = 7.243 \times 10^{21} \times 10^{-3}/296 = 2.4 \times 10^{16}\ \text{molec. dm}^{-3}$$

For $p = 10^{-6}$ mbar

$$n = 7.243 \times 10^{21} \times 10^{-6}/296 = 2.4 \times 10^{13} \text{ molec. dm}^{-3}$$

For $p = 10^{-9}$ mbar

$$n = 7.243 \times 10^{21} \times 10^{-9}/296 = 2.4 \times 10^{10} \text{ molec. dm}^{-3}$$

(ii) Calculate the ratio $A_s/V_s$

$$A_s = \pi d^2$$

$$V_s = \pi d^3/6$$

$$A_s/V_s = 6/d$$

Substitute $d$ (the diameter of the sphere) = 5 dm.
Thus

$$A_s/V_s = 6/5 \text{ dm} = 1.2 \text{ dm}^{-1}$$

(iii) Calculate

$$r = (n_m/n)(A_s/V_s) \qquad \text{dimensionless}$$

For $p = 1.0$ mbar

$$r = \left[(8.16 \times 10^{16} \text{ molec. dm}^{-2})/(2.4 \times 10^{19} \text{ molec. dm}^{-3})\right] \times 1.2 \text{ dm}^{-1}$$

$$= 4.08 \times 10^{-3}$$

For $p = 10^{-3}$ mbar

$$r = \left[(8.16 \times 10^{16} \text{ molec. dm}^{-2})/(2.4 \times 10^{16} \text{ molec. dm}^{-3})\right] \times 1.2 \text{ dm}^{-1}$$

$$= 4.08$$

For $p = 10^{-6}$ mbar

$$r = \left[(8.16 \times 10^{16} \text{ molec. dm}^{-2})/(2.4 \times 10^{13} \text{ molec. dm}^{-3})\right] \times 1.2 \text{ dm}^{-1}$$

$$= 4.08 \times 10^{3}$$

For $p = 10^{-9}$ mbar

$$r = \left[(8.16 \times 10^{16} \text{ molec. dm}^{-2})/(2.4 \times 10^{10} \text{ molec. dm}^{-3})\right] \times 1.2 \text{ dm}^{-1}$$

$$= 4.08 \times 10^{6}$$

(*Note*: The lower the pressure, the higher the number of molecules in a monolayer on the surface of the sphere compared to the number of molecules in the volume of the sphere.)

**Exercise 3.12**

Determine the mean free path and the time required to form a monolayer for nitrogen molecules at $10^{-6}$ mbar and $T = 23°C$.

**Solution 3.12**

Use Eqs. 3.23d and 3.22c:

$$\lambda = 3.108 \times 10^{-20} T/(p\delta_m^2) \qquad \text{cm}$$

$$t_m = 3.794 \times 10^{-23} (TM)^{1/2}/(p\delta_m^2) \qquad \text{s}$$

Substitute

$p = 10^{-6}$ mbar
$T = 273 + 23 = 296$ K
$\delta_m = 3.78 \times 10^{-8}$ cm (see Table 3.6)
$M = 28$ g mol$^{-1}$ (see Table 1.2)

Thus

$$\lambda = 3.108 \times 10^{-20} \times 296/(10^{-6} \times (3.78 \times 10^{-8})^2$$

$$= 6439 \text{ cm} = 64.39 \text{ m}$$

$$t_m = 3.794 \times 10^{-23} \times (296 \times 28)^{1/2}/10^{-6} \times (3.78 \times 10^{-8})^2 = 2.42 \text{ s}$$

**Exercise 3.13**

Two chambers containing the same species of gas are connected by a tube of diameter $d$. Describe what happens when (a) the temperature of the gas in the chambers varies and the mean free path of gas molecules is extremely large compared to the diameter of the connecting tube and (b) the temperature of the gas in the chambers reaches equilibrium and the mean free path of the gas molecules is very small compared to the diameter of the connecting tube.

**Solution 3.13**

Let subscripts 1 and 2 to physical quantities identify the state of the gas in the two chambers.

(a) The mass of gas incident on unit area of the connecting tube per unit time is (Eq. 3.14b)

$$G_1 = \rho_1 v_{a1}/4$$

$$G_2 = \rho_2 v_{a2}/4$$

Since $\lambda \gg d$, molecular collisions are very rare near the opening of the connecting tube. Thus $G_1 = G_2$ so that

$$\rho_1 v_{a1} = \rho_2 v_{a2}$$

On substituting the expression of $v_a$ by Eq. 3.2

$$v_a = (8RT/\pi M)^{1/2}$$

There results

$$\rho_1(8RT_1/\pi M_1)^{1/2} = \rho_2(8RT_2/\pi M_1)^{1/2}$$

$$\rho_1 T_1^{1/2} = \rho_2 T_2^{1/2}$$

$$\rho_1/\rho_2 = (T_2/T_1)^{1/2}$$

Substituting $\rho$ by Eq. 1.14a

$$\rho = pM/RT$$

There results

$$(p_1M_1/RT_1)/(p_2M_2/RT_2) = (T_2/T_1)^{1/2}$$

and

$$p_1/p_2 = (T_1/T_2)^{1/2}$$

(b) When $\lambda \ll d$ and the gas in chambers 1 and 2 reaches thermal equilibrium such that $T_1 = T_2$, the collision frequency between molecules in the neighborhood of the orifice of the connecting tube causes a mass motion of the gas through the orifice until $p_1 = p_2$.

**Exercise 3.14**

A capacitance manometer is attached by a tube of narrow diameter to a vacuum system whose temperature is maintained constant at $T = 23°C$. The transducer of the capacitance manometer is maintained at $T = 35°C$. When the pressure in the vacuum system is less than $\sim$ 100 Pa, the manometer reads a pressure higher than the actual pressure in the system. Calculate the pressure read by the capacitance manometer, due to the thermal transpiration effect.

**Solution 3.14**

Use Eq. 3.43 written as

$$p_2 = p_1(T_2/T_1)^{1/2}$$

where $p_2$ and $T_2$ are the pressure and temperature of the capacitance manometer transducer whereas $p_1$ and $T_1$ are respectively the pressure and temperature of the vacuum system.
Substitute

$$T_2 = 273 + 35 = 208 \text{ K}$$

$$T_1 = 273 + 23 = 296 \text{ K}$$

Thus

$$p_2 = p_1(308 \text{ K}/296 \text{ K})^{1/2} = 1.02\ p_1$$

## Exercise 3.15

Nitrogen at a pressure $p_1$ is supplied via a cold trap at $T = -78°C$ to a vacuum system maintained at $T = 23°C$. A McLeod manometer, at the same temperature as the system, measures the pressure $p_2$ in the system. Calculate the actual pressure of nitrogen in the system.

## Solution 3.15

Owing to the difference between the temperature of the cold trap $T_1 = -78°C$ and the temperature of the McLeod manometer $T_2 = 23°C$, the actual pressure $p_1$ should be calculated by taking into account the thermal transpiration effect. Use Eq. 3.43 written as

$$p_1 = p_2(T_1/T_2)$$

Substitute

$$T_1 = 273 - 78 = 195 \text{ K}$$

$$T_2 = 273 + 23 = 296 \text{ K}$$

Thus

$$p_1 = p_2(195 \text{ K}/296 \text{ K})^{1/2} = 0.811 p_2$$

## Exercise 3.16

The calibration chamber of a single-stage expansion system is connected to a capacitance manometer by a tube of 4-mm inner diameter (ID). The temperatures, maintained constant, are 20°C at the calibration system and 40°C at the transducer of the capacitance manometer. Three successive pressure points are generated into the calibration chamber, using nitrogen, for each decade within the range $10^2$ to $10^{-1}$ Pa. The value of each generated pressure point ($p_g$) measured by a primary standard and the readings ($p_{cap}$) of the capacitance manometer corresponding to the pressure generated into the calibration chamber are given in Table 3A (for clarity of explanation, the uncertainties of the measurement at a 95% confidence level have been omitted).

**Table 3A**

| Pressure | | | (Pa) | | | |
|---|---|---|---|---|---|---|
| $p_c$, measured | $6.70 \times 10^2$ | $2.83 \times 10^2$ | $1.25 \times 10^2$ | $6.52 \times 10$ | $2.74 \times 10$ | $1.31 \times 10$ |
| $p_g$, calculated | $6.68 \times 10^2$ | $2.82 \times 10^2$ | $1.24 \times 10^2$ | $6.50 \times 10$ | $2.73 \times 10$ | $1.30 \times 10$ |
| $p_c$, measured | 6.85 | 2.81 | 1.33 | $6.63 \times 10^{-1}$ | $2.75 \times 10^{-1}$ | $1.22 \times 10^{-1}$ |
| $p_g$, calculated | 6.74 | 2.75 | 1.29 | $6.44 \times 10^{-1}$ | $2.66 \times 10^{-1}$ | $1.18 \times 10^{-1}$ |

**Table 3B**

| Pressure | (Pa) | | | | | |
|---|---|---|---|---|---|---|
| $p_c$, measured | $6.70 \times 10^2$ | $2.83 \times 10^2$ | $1.25 \times 10^2$ | $6.52 \times 10$ | $2.74 \times 10$ | $1.31 \times 10$ |
| $p_c/p_g$, calculated | 1.0000 | 1.0000 | 1.0002 | 1.0005 | 1.0024 | 1.0068 |
| $p_c$, measured | 6.85 | 2.81 | 1.33 | $6.63 \times 10^{-1}$ | $2.75 \times 10^{-1}$ | $1.22 \times 10^{-1}$ |
| $p_c/p_g$, calculated | 1.0126 | 1.0202 | 1.0253 | 1.0277 | 1.0304 | 1.0320 |

Plot the calibration curve of the capacitance manometer, accounting for thermal transpiration calculated according to Takaishi and Sensui (1963) method.

**Solution 3.16**

The plot of the capacitance manometer output corrected for thermal transpiration involves the evaluation of this effect for each point of pressure generated ($p_g$) and subsequently measured ($p_{cap}$). To avoid redundancy, calculations will address a single pressure point. The result of similar calculations for the other pressure points is shown in Table 3B.

**Solution 3.16**

From Eq. 3.43a

$$p_2/p_1 = \left[A^*(X/\overline{T})^2 + B^*(X/\overline{T}) + C^*(X/\overline{T})^{1/2} + (T_2/T_1)^{1/2}\right] \Big/ \left[A^*(X/\overline{T})^2 + B^*(X/\overline{T}) + C^*(X/\overline{T})^{1/2} + 1\right]$$

$$\overline{T} = (T_1 + T_2)/2 \qquad \text{K}$$

$$X = pd \qquad \text{Pa m}$$

Substitute

$A^* = 67.5 \times 10^6\ (\text{K}^2\text{Pa}^{-2}\text{m}^{-2})$ (see Table 3.17)
$B^* = 7.5 \times 10^3\ (\text{K Pa}^{-1}\ \text{m}^{-1})$ (see Table 3.17)
$C^* = 38.34\ [\text{K}^{1/2}\text{Pa}^{-(1/2)}\text{m}^{-(1/2)}]$ (see Table 3.17)
$p_2 = p_g 2.75 \times 10^{-1}$ Pa
$p_1 = p_{cap} = 2.66 \times 10^{-1}$ Pa
$T_2 = T_g = 273 + 20 = 293$ K
$T_1 = T_{cap} = 273 + 40 = 313$ K
$d = 4 \times 10^{-3}$ m

Calculate

$$\chi = 2.75 \times 10^{-1}\ \text{Pa} \times 4 \times 10^{-3}\ \text{m} = 1.1 \times 10^{-3}\ \text{Pa m}$$

$$\overline{T} = (293\ \text{K} + 313\ \text{K})/2 = 303\ \text{K}$$

$$(T_2/T_1)^{1/2} = (T_g/T_{cap})^{1/2} = (293\ \text{K}/313\ \text{K})^{1/2} = 0.9675$$

$$A^*(\chi/\overline{T})^2 = 67.5 \times 10^6\ \text{K}^2\ \text{Pa}^{-2}\ \text{m}^{-2} \times \left[(1.1 \times 10^{-3}\ \text{Pa m})/(303\ \text{K})\right]^2$$

$$= 8.896 \times 10^{-4}$$

$$B^*(\chi/\overline{T}) = 7.5 \times 10^3\ \text{K Pa}^{-1}\ \text{m}^{-1} \times 1.1 \times 10^{-3}\ \text{Pa m}/301\ \text{K}$$

$$= 2.723 \times 10^{-2}$$

$$C^*(\chi/\overline{T})^{1/2} = 38.34\ \text{K}^{1/2}\ \text{Pa}^{-1/2}\ \text{m}^{-1/2}$$

$$\times\left[(1.1 \times 10^{-3}\ \text{Pa m})/(303\ \text{K})\right]^{1/2}$$

$$= 7.305 \times 10^{-2}$$

$$p_g/p_c = (8.896 \times 10^{-4} + 2.273 \times 10^{-2}$$

$$+7.305 \times 10^{-2} + 0.9675)/$$

$$(8.896 \times 10^{-4} + 2.273 \times 10^{-2} + 7.305 \times 10^{-2} + 1)$$

$$= 1.0686696/1.1011696 = 0.9704859$$

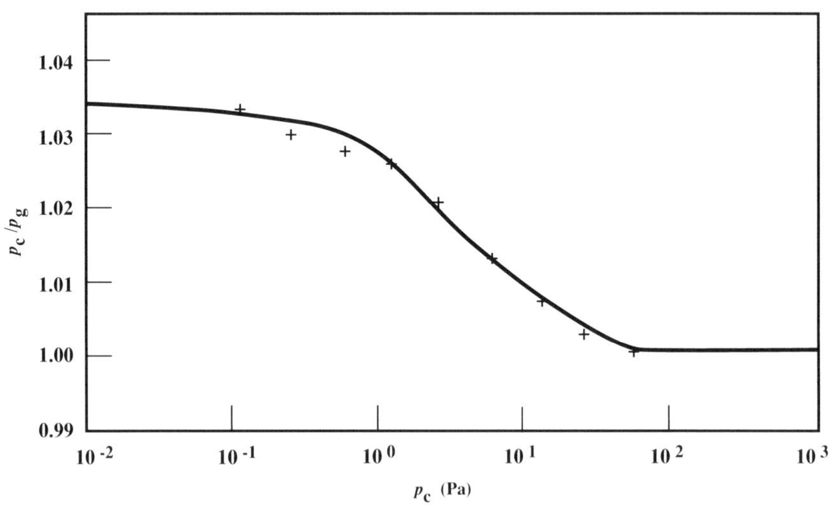


Figure 3A

The error owing to the thermal transpiration effect on the readings of the capacitance manometer is

$$p_c/p_g = 1/0.9704859 = 1.0304117 \quad \text{(i.e, 3.04\%)}$$

Similar calculations for all the values listed in Table 3A are shown in Table 3B.

A plot of the ratio $p_c/p_g$ measured (solid line in Fig. 3A) versus the same ratio calculated (+) shows that the experimental and calculated data differ by no more than 0.5%.

## References


Armdur, I., and Guildner, L. A. (1957), *J. Am. Chem. Soc.*, *79*, 311.

Berman, A. (1985), *Total Pressure Measurements in Vacuum Technology*, Academic Press, New York.

Buckmann, S. J., Gascoigne, J., and Roberts, K. B. (1984), *J. Vac. Sci. Technol. A2*, *4*, 1599.

CRC (1986–1987), *Handbook of Chemistry and Physics*, 67th ed., R. C. Weast (ed.), CRC Press, Boca Raton, Fla.

De Porter, L. G., and Searcy, A. W. (1963), *J. Chem. Phys.*, *39*, 925.

Diels, K., and Jaeckel, R. (1958), *Leybold Vakuum—Taschenbuch für Laboratorium and Betrieb*, Springer-Verlag, Berlin.

Dushman, S. (1966), *Scientific Foundations of Vacuum Technique*, 2nd ed., J. M. Lafferty (ed.), Wiley, New York.

Edwards, D. Jr. (1978), *J. Vac. Sci. Technol.*, *16*, 758.

Eggelton, A. E., and Tompkins, F. C. (1952), *Trans. Farad. Soc.*, *48*, 738.

Jitschin, W., and Röhl, P. (1987), *J. Vac. Sci. Technol. A5*, *3*, 372.

Knudsen, M. (1910), *Ann. Physik*, *31*, 210, 633.

Poulter, K. F., Rodgers, M. J., Nash, P. J., Thompson, T. J., and Perkin, M. P. (1983), *Vacuum*, *33*, 6, 31.

Present, R. D. (1958), *Kinetic Theory of Gases*, McGraw-Hill, New York.

Roth, A. (1982), *Vacuum Technology*, North-Holland, Amsterdam.

Takaishi, T., and Sensui, Y. (1963), *Trans. Faraday Soc.*, *59*, 2503.

Thomas, L. B., and Brown, R. E. (1950), *J. Chem. Phys.*, *18*, 1367.

Thomas, L. B., and Schofeild, E. B. (1955), *J. Chem. Phys.*, *23*, 861.

# 4

# Gas Flow in Components and Vacuum Systems

## 4.1 *Quantity of Gas*

The quantity of gas can be measured in one of the following ways:

(a) In pressure–volume units at a specified temperature:

$$a_{pv} = pV \quad \text{at} \quad T \qquad (\text{e.g., Pa m}^3) \tag{4.1}$$

(b) In moles:

$$a_{mol} = W_t/M \qquad (\text{mol}) \tag{4.1a}$$

$$= a_{pv}/RT \tag{4.1b}$$

(c) In molecules:

$$a_m = W_t/m \qquad (\text{molec.}) \tag{4.1c}$$

$$= a_{pv}/kT \tag{4.1d}$$

(d) In mass units:

$$a_{kg} = W_t \qquad (\text{kg}) \tag{4.1e}$$

$$= a_{pv} M/RT \tag{4.1f}$$

(*Note*: Use consistent units for $p$, $V$, and $R$.)

## 4.2 *Molar Flow Rate of Gas*

### 4.2.1 Single Species of Gas

*a. Instantaneous.*

$$q_{mol} = d/dt(a_{mol}) \qquad (mol\ s^{-1}) \tag{4.2}$$

$$= d/dt(W_t/M) \tag{4.2a}$$

$$= (1/RT)\ d/dt(pV) \tag{4.2b}$$

Equation 4.2 becomes
For $p$ = const.:

$$q_{mol} = (p/RT)d/dt(V) \qquad (mol\ s^{-1}) \tag{4.2c}$$

For $V$ = const.:

$$q_{mol} = (V/RT)d/dt(p) \qquad (mol\ s^{-1}) \tag{4.2d}$$

*b. Average, over a time interval $\Delta t$.*

$$q_{mol} = d/dt\left(\overline{W}_t/\overline{M}\right) \qquad (mol\ s^{-1}) \tag{4.3}$$

$$= (p_1V_1/RT_1 - p_2V_2/RT_2)/\Delta t \tag{4.3a}$$

where subscripts 1 and 2 to pressure $p$ and volume $V$ refer to measurements at the beginning and the end of the interval $\Delta t$, respectively.

### 4.2.2 Mixture of Gases

Calculations should be made as shown in Section 4.2.1 using the partial pressure and molar mass of each gas species in the mixture.

## 4.3 *Conversion of Molar Flow Rates to Molecular or Mass Flow Rates*

$$q_m = q_{mol} N_A \qquad (molec.\ s^{-1}) \tag{4.4}$$

$$q_{kg} = q_{mol} M \qquad (kg\ s^{-1}) \tag{4.5}$$

## 4.4 Throughput

$$Q = d/dt(a_{pv}) \qquad (\text{e.g., Pa m}^3\ \text{s}^{-1}\ \text{at}\ T) \tag{4.6}$$

$$= p\,dV/dt \text{ for } p = \text{const.} \tag{4.6a}$$

$$= pV' \tag{4.6b}$$

$$= q_m RT/N_A \tag{4.6c}$$

$$= q_{kg} RT/M \tag{4.6d}$$

where $V' = dV/dt$ at a specified temperature $T$.

Minor variations in temperature have no effect on the throughput. However, in some applications such as leak detection, it is desirable to specify the temperature since that would enable comparison of throughputs on an equal basis. *To avoid ambiquities, the temperature at which throughput is measured must be clearly specified after the units, e.g.,* 5.0 × $10^{-3}$ *Pa m*$^3$ *s*$^{-1}$*, at* $T = 23°C$.

The SI unit for throughput is W. It expresses the power required to move gas, at constant flow rate, across a specified cross section of the vacuum system. Thus

$$1\ \text{Pa m}^3\ \text{s}^{-1} = 1\ \text{N m}^{-2}\ \text{m}^3\ \text{s}^{-1} = 1\ \text{N m s}^{-1} = 1\ \text{J s}^{-1} = 1\ \text{W}$$

(*Note*: Use consistent units for $p$, $R$, and $T$.)

## 4.5 Conversion of Throughputs

(a) From a specified temperature $T_1$ to another temperature $T_2$:

$$Q_1 \quad (\text{at } T_1) \times T_2/T_1 = Q_2 \quad (\text{at } T_2) \tag{4.7}$$

(b) From one species of gas 1 to another species of gas 2 at $T$ = const.:

$$Q_{11} = Q_{12}(\eta_2/\eta_1) \qquad (\text{laminar flow}) \tag{4.8}$$

$$Q_{m1} = Q_{m2}(M_2/M_1)^{1/2} \qquad (\text{molecular flow}) \tag{4.9}$$

(c) From a specified temperature $T_g$ and species of gas $M_g$ to a mass flow rate:

$$q_{kg} = Q_{Mg\,Tg} \times M_g/RT_g \qquad (\text{kg s}^{-1}) \tag{4.10}$$

The conversion factors for throughputs and leak rates at $T = 0°C$ in different systems of units are listed in Table 4.1. For conversion factors accurate within 5 digits after the decimal point, see Moss (1987).

**Table 4.1**

Conversion Factors for Throughput and Leak Rates at $T = 0°C$

| | Pa $m^3$ $s^{-1}$ or W | mbar liters $s^{-1}$ | Torr liters $s^{-1}$ (1 $\mu$s) | $\mu$liters $s^{-1}$ (1 $\mu$s) | $\mu ft^3$ $s^{-1}$ | $cm^{3a}$ $s^{-1}$ STP | mol $s^{-1}$ air | molec. $s^{-1}$ air | mg $s^{-1}$ air |
|---|---|---|---|---|---|---|---|---|---|
| Pa $m^3$ $s^{-1}$ or W | 1.0 | $1.0 \times 10^1$ | $7.5 \times 10^0$ | $7.5 \times 10^3$ | $2.65 \times 10^2$ | $9.87 \times 10^0$ | $4.40 \times 10^{-4}$ | $2.65 \times 10^{20}$ | $1.28 \times 10$ |
| mbar liters $s^{-1}$ | $1.0 \times 10^{-1}$ | 1.0 | $7.5 \times 10^{-1}$ | $7.5 \times 10^2$ | $2.65 \times 10^1$ | $9.87 \times 10^{-1}$ | $4.40 \times 10^{-5}$ | $2.69 \times 10^{19}$ | $1.28 \times 10^0$ |
| Torr liters $s^{-1}$ | $1.33 \times 10^{-1}$ | $1.33 \times 10^0$ | 1.0 | $1.0 \times 10^3$ | $3.53 \times 10^1$ | $1.32 \times 10^0$ | $5.87 \times 10^{-5}$ | $3.54 \times 10^{19}$ | $1.70 \times 10^0$ |
| $\mu$ liters $s^{-1}$ (1 $\mu$s) | $1.33 \times 10^{-4}$ | $1.33 \times 10^{-3}$ | 1.0 | $3.53 \times 10^{-2}$ | $1.32 \times 10^{-3}$ | $5.87 \times 10^{-8}$ | $3.54 \times 10^{16}$ | $1.70 \times 10^{-3}$ | |
| $\mu ft^3$ $s^{-1}$ | $3.78 \times 10^{-3}$ | $3.78 \times 10^{-2}$ | $2.83 \times 10^{-2}$ | $2.83 \times 10^1$ | 1.0 | $3.73 \times 10^{-2}$ | $1.66 \times 10^{-6}$ | $1.0 \times 10^{18}$ | $4.18 \times 10^{-2}$ |
| $cm^3$ $s^{-1}$ STP[a] | $1.01 \times 10^{-1}$ | $1.01 \times 10^0$ | $7.60 \times 10^{-1}$ | $7.60 \times 10^2$ | $9.67 \times 10^4$ | 1.0 | $4.46 \times 10^{-5}$ | $2.69 \times 10^{19}$ | $1.29 \times 10^0$ |
| mol $s^{-1}$ air | $2.27 \times 10^3$ | $2.27 \times 10^4$ | $1.70 \times 10^4$ | $1.70 \times 10^7$ | $6.02 \times 10^5$ | $3.24 \times 10^4$ | 1.0 | $6.02 \times 10^{23}$ | $2.89 \times 10^4$ |
| molec. $s^{-1}$ air | $3.77 \times 10^{-21}$ | $3.77 \times 10^{-20}$ | $2.83 \times 10^{-20}$ | $2.83 \times 10^{-17}$ | $9.99 \times 10^{-19}$ | $3.72 \times 10^{-20}$ | $1.66 \times 10^{-24}$ | 1.0 | $4.80 \times 10^{-20}$ |
| mg $s^{-1}$ air | $7.84 \times 10^{-2}$ | $7.84 \times 10^{-1}$ | $5.88 \times 10^{-1}$ | $5.88 \times 10^2$ | $2.08 \times 10^1$ | 7.74h $10^{-1}$ | $3.45 \times 10^{-5}$ | $2.08 \times 10^{19}$ | 1.0 |

[a]Frequently expressed as standard cubic centimeters (cc) $s^{-1}$ or $cm^3$ atm $s^{-1}$ at STP.

**Table 4.2**
Gas Flow Regimes in Pipes and Vacuum Components When Gas Pressure Is Reduced from Atmosphere to the Lowest Attainable Pressure

| Gas density | Gas state | Flow regime | Observations |
|---|---|---|---|
| High | Viscous | Turbulent or laminar | The mean free path of gas molecules is small compared to the diameter of the pipe; the gas behaves like a continuous fluid |
| Medium | Transition between viscous and tenuous | Transition | |
| Low | Tenuous | Molecular | The mean free path of gas molecules is extremely large compared to the diameter of the pipe; a molecule leaving the wall of the pipe after a collision has equal probabilities of continuing its former direction or returning toward its initial starting point |

## 4.6 *Gas Flow Regimes*

The gas flow regimes that develop in pipes, vacuum components, and vacuum systems when gas pressure is reduced from atmosphere to the lowest attainable value are listed in Table 4.2. These regimes depend on the size of the vacuum pipes (components), the gas species, and temperature.

### 4.6.1 Quantifying Criteria of Gas Flow

*a. The Reynolds number.* The flow of gas through a pipe, retaining the characteristics of a continuous fluid, may be characterized and quantified by the Reynolds number Re

$$\mathrm{Re} = dv\rho/\eta \qquad \text{(dimensionless)} \tag{4.11}$$

where $d$ = diameter of the pipe through which gas flows
$v$ = gas velocity along the portion of the pipe in question
$\rho$ = density of gas at the temperature of the flow
$\eta$ = viscosity of gas at the temperature of the flow

(*Note*: Use consistent units for $d$, $v$, $\rho$, and $\eta$.)

For a pipe of circular cross section and diameter $d$ the velocity $v$ is expressed as

$$v = V'/(\pi d^2/4) \qquad (\text{e.g., m h}^{-1}) \tag{4.12}$$

$$= 4Q/(\pi d^2 p_i) \tag{4.12a}$$

where $V'$ = volume rate of gas flow at a specified temperature
$p_i$ = gas pressure measured at that plane
$Q$ = gas throughput, at the temperature specified for $V'$, flowing across the plane where $p_i$ is measured
$d$ = diameter of the pipe

Consequently, for

$$\text{Re} > 2200 \text{ (approximately) } \textit{the flow is turbulent} \tag{4.13}$$

$$\text{Re} < 1200 \text{ (approximately) } \textit{the flow in laminar} \tag{4.14}$$

*b. The throughput–Reynolds number relationship.*

$$Q = (\text{Re}d)(\pi RT\eta/4M) \qquad (\text{e.g., Torr liters s}^{-1}) \tag{4.15}$$

$$Q = (\text{Re}d)(\pi kT\eta/4m) \tag{4.15a}$$

On substituting in Eq. 4.15 the values for Re for turbulent (Eq. 4.13) or laminar flow, as well as the values of $\eta$ and $M$ both for air at $T = 23°\text{C}$ and $R$, the relationship between the gas throughput $Q$ and the diameter $d$ of the pipe becomes, in different systems of units

(i) Turbulent flow of air at $T = 23°\text{C}$ (subscript t to $Q$):

$$Q_t > 2.62 \times 10^5 d \qquad (\text{Pa m}^3\ \text{s}^{-1}) \tag{4.16}$$

with $d$ (m),

$$Q_t > 262d \qquad (\text{mbar liters s}^{-1}) \tag{4.17}$$

with $d$ (cm), and

$$Q_t > 196d \qquad (\text{Torr liters s}^{-1}) \tag{4.18}$$

with $d$ (cm).

Turbulent flow of gas occurs for only a short period of time at the beginning of the evacuation of a vessel from atmospheric pressure. It has been found experimentally that the turbulent flow is approximately proportional to the root of the pressure gradient.

(ii) Laminar flow of air at $T + 23°\text{C}$ (subscript $l$ to $Q$)

$$Q_l < 1.43 \times 10^5 d \qquad (\text{Pa m}^3\ \text{s}^{-1}) \tag{4.19}$$

with $d$ (m),

$$Q_l < 143d \qquad (\text{mbar liters s}^{-1}) \tag{4.20}$$

with $d$ (cm), and

$$Q < 107d \qquad (\text{Torr liters s}^{-1}) \tag{4.21}$$

with $d$ (cm).

*c. The Knudsen number.* Knudsen (1910) found experimentally that the number $K' = \lambda/d$ (see Eq. 3.26) can supply information concerning the gas-flow regime. Thus, when

$$K' < 0.01 \qquad \text{the flow is laminar (turbulent)} \tag{4.22}$$

$$K' > 1.0 \qquad \text{the flow is molecular} \tag{4.23}$$

$$1.0 > K' > 0.01 \qquad \text{the flow is in the transition range} \tag{4.24}$$

*d. Relationship between average pressure $\bar{p}$ within pipe and pipe diameter d.* If $p_1$ and $p_2$ are the pressures of gas at the inlet and outlet of the pipe, respectively, the average pressure is

$$\bar{p} = (p_1 + p_2)/2 \qquad \text{(e.g., Pa)} \tag{4.25}$$

Then the flow regimes are expressed as follows:

(i) Laminar (or turbulent) flow of air at $T = 23°C$:

$$\bar{p}d > 0.65 \qquad \text{(e.g., mbar cm)} \tag{4.26}$$

with $\bar{p}$ (Pa or mbar), $d$ (m or cm), and

$$\bar{p}d > 0.49 \qquad \text{(Torr cm)} \tag{4.27}$$

with $\bar{p}$ (Torr) and, $d$ (cm).

(ii) Molecular flow of air at $T = 23°C$

$$\bar{p}d < 6.5 \times 10^{-3} \qquad \text{(e.g., Pa m)} \tag{4.28}$$

with $\bar{p}$ (Pa or mbar), $d$ (m or cm) and

$$\bar{p}d < 4.9 \times 10^{-3} \qquad \text{(Torr cm)} \tag{4.29}$$

with $\bar{p}$ (Torr), and $d$ (cm).

(*Note*: Use consistent units for $\bar{p}$ and $d$.)

The criteria for checking the air-flow regime at $T = 23°C$ in pipes of circular cross section, according to Eqs. 4.13–4.29, are summarized in Table 4.3.

## 4.7 Impedance to Gas Flow

Under *steady-state conservative conditions* ($T$ = const., $Q$ = const., absence of gas load) of gas flow through a vacuum system, any component of the system opposes a resistance to the gas passage, known as *impedance*. The expression of

**Table 4.3**
Criteria for Checking the Gas Flow Regime in a Pipe of Circular Cross Section

| Criteria | Flow regime | | | |
|---|---|---|---|---|
| | Viscous | | | |
| | Turbulent | Laminar | Transition | Molecular |
| a. Reynolds number Re (dimensionless), Eqs. 4.13 and 4.14 | Re > 2200 | Re < 1200 | — | — |
| b. Throughput–Reynolds number relationship for air at $T = 23°C$, Eqs. 4.16–4.21 | | | | |
| $Q$ (Pa m$^3$ s$^{-1}$) $d$ (m) | $Q_t > 2.62 \times 10^5 d$ | $Q_l < 1.43 \times 10^5 d$ | — | — |
| $Q$ (mbar s$^{-1}$) $d$ (cm) | $Q_t > 262d$ | $Q_l < 143d$ | — | — |
| $Q$ (Torr s$^{-1}$) $d$ (cm) | $Q_t > 196d$ | $Q_l < 107d$ | — | — |
| c. Knudsen number $K'$ (dimensionless), Eqs. 4.22–4.24 | $K' < 0.01$ | | $1 > K' > 1.0 \times 10^{-2}$ | $K' > 1.0$ |
| d. Relationship between average pressure $\bar{p}$ within pipe and diameter of that pipe, for air at $T = 23°C$, Eqs. 4.25–4.29 | | | | |
| $\bar{p}$ (Pa or mbar) $d$ (m or cm) | $\bar{p}d > 0.65$ | | $6.5 \times 10^{-1} > \bar{p}d > 6.5 \times 10^{-3}$ | $\bar{p}d < 6.5 \times 10^{-3}$ |
| $\bar{p}$ (Torr) $d$ (cm) | $\bar{p}d > 0.49$ | | $4.9 \times 10^{-1} > \bar{p}d > 4.9 \times 10^{-3}$ | $\bar{p}d < 4.9 \times 10^{-3}$ |

this impedance (Diels and Jaeckel, 1958, p. 54) is

$$Z = (1/q_{kg})(p_1 - p_2)M/RT \qquad (\text{e.g., s}^{-1}) \qquad (4.30)$$

where subscripts 1 and 2 refer to the pressure at the entrance (upstream) and exit (downstream) from the component, respectively.

## 4.8 Conductance to Gas Flow

$$C = 1/Z \qquad (\text{e.g., m}^3\ \text{h}^{-1}) \qquad (4.31)$$

$$= (q_{kg}RT/M)/(p_1 - p_2) \qquad (4.31a)$$

$$= Q/(p_1 - p_2) \qquad (4.31b)$$

Table 4.4
Conversion Factors for Conductance and Pumping Speed—Volume Rate of Flow

| | liters $s^{-1}$ | liters $min^{-1}$ | $ft^3\ min^{-1}$ | $m^3\ h^{-1}$ | $cm^3\ s^{-1}$ |
|---|---|---|---|---|---|
| liters $s^{-1}$ | 1.0 | 6.0 | 2.12 | 3.6 | 1000 |
| liters $min^{-1}$ | 0.0167 | 1.0 | 0.0353 | 0.060 | 16.67 |
| $ft^3\ min^{-1}$ | 0.47 | 28.32 | 1.0 | 1.699 | 471.95 |
| $m^3\ h^{-1}$ | 0.278 | 16.67 | 0.589 | 1.0 | 277.8 |
| $cm^3\ s^{-1}$ | 0.001 | 0.06 | 0.0021 | 0.0036 | 1.0 |

The impedance Z (Eq. (4.30) and conductance C (Eq. 4.31a) of a vacuum component to gas flow vary with gas species M and the temperature $T$ of gas, and, except in molecular flow, with gas pressure. The values of Z and C depend on the gas flow regime and the geometry of the component.

The conversion factors for conductance in different systems of units are listed in Table 4.4.

## 4.9 *Comparison between an Electrical and Vacuum Circuits*

If a simple electrical circuit, consisting of a continuous electrical current source connected to a wire resistance, is compared to a simple vacuum circuit, consisting of a vacuum pump that evacuates through a pipe or a vessel having a constant leak, the following analogy can be established:

Electric source: vacuum pump
Electric current: throughput
Wire resistance: Impedance of the pipe

It must, however, be kept in mind that this comparison cannot be carried too far.

Wilson (1987) has modeled a vacuum system as an electrical network. His method includes the capability to model outgassing and to obtain the response of a vacuum system to transient and time-varying loads.

## 4.10 *Impedance and Conductance of Interconnected Vacuum Components*

In this paragraph, subscripts 1, 2, . . . , $n$ and T to Z or C refer to the individual or total impedance (conductance) of the components, respectively.

### 4.10.1 Vacuum Components Connected in Series

$$Z_T = Z_1 + Z_2 + \cdots + Z_n \quad (\text{e.g., s liter}^{-1}) \tag{4.32}$$

$$Z_{1,2} = Z_1 + Z_2 \tag{4.32a}$$

$$1/C_T = 1/C_1 + 1/C_2 + \cdots + 1/C_n \quad (\text{e.g., s liter}^{-1}) \tag{4.33}$$

$$C_{1,2} = C_1C_2/(C_1 + C_2) \quad (\text{e.g., liters s}^{-1}) \tag{4.33a}$$

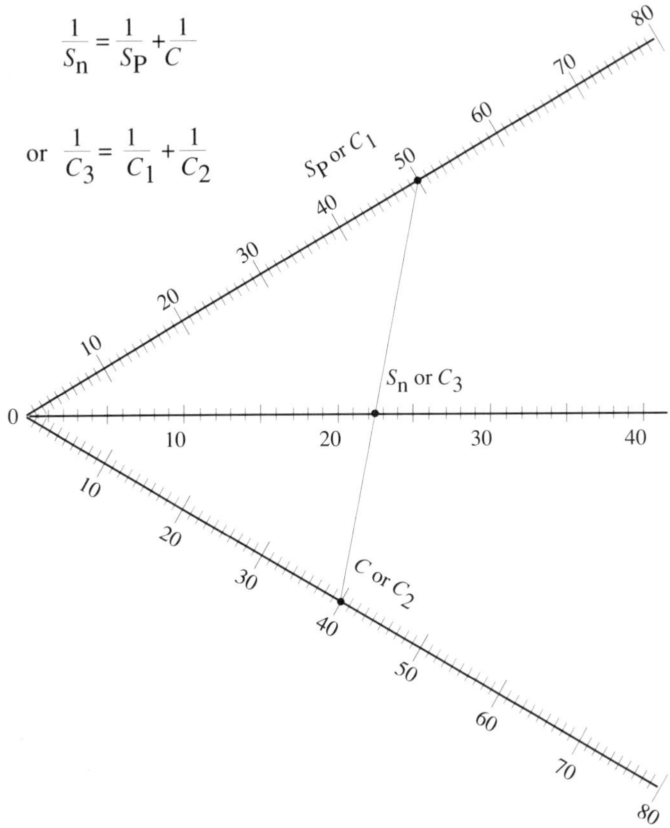


**Figure 4.1** Nomograph for evaluating either the conductance C, the pumping speed $S_P$ at the port of the vacuum pump, or the net pumping speed $S_n$ at the port of a vacuum vessel when two parameters (e.g., C and $S_P$) are known. If necessary, the range of the scales must be multiplied by the same factor. Use consistent units for all scales. [After Delafosse and Mongodin (1961).]

### 4.10.2 Vacuum Components Connected in Parallel

$$1/Z_T = 1/Z_1 + 1/Z_2 + \cdots + 1/Z_n \qquad (\text{e.g., m}^3\ \text{h}^{-1}) \quad (4.34)$$

$$Z_{1,2} = Z_1 Z_2/(1/Z_1 + 1/Z_2) \qquad (\text{e.g., h m}^{-3}) \quad (4.34a)$$

$$C_T = C_1 + C_2 + \cdots + C_n \qquad (\text{e.g., m}^3\ \text{h}^{-1}) \quad (4.35)$$

$$C_{1,2} = C_1 + C_2 \qquad (\text{e.g., m}^3\ \text{h}^{-1}) \quad (4.35a)$$

For certain conditions of molecular flow in short pipes and components Eqs. 4.32–4.35 are not accurate since end effects are not considered.

The conductance of an assembly of two vacuum elements connected in series can be evaluated using the nomograph in Fig. 4.1.

## *4.11 Conversion of Conductances for Cylindrical Pipes of Uniform Cross Section*

A pipe of length $L$ and diameter $d$ is qualified as "long" when $L > 20d$.

### 4.11.1 Conversion for Laminar Flow

Subscript l at $T$ = const., from one species of gas to another:

$$C_{l1}/C_{l2} = \eta_2/\eta_1 \quad (4.36)$$

### 4.11.2 Conversion for Molecular Flow

Subscript m, from a temperature $T_1$ and species of gas $M_1$ to $T_2$ and $M_2$:

$$C_{m1}/C_{m2} = (M_2/M_1)^{1/2}(T_1/T_2)^{1/2} \quad (4.37)$$

## *4.12 Pumping Speed—Volume Rate of Flow—Permanent Regime*

Molecules or atoms are said to have been "pumped" if removed from active circulation within a system for a *long period of time* (Lawrence and Lamont, 1973, p. 231).

### 4.12.1 Pumping Speed of a Vacuum Component or a Vacuum Pump

$$S = dV/dt \text{ at pressure } p_s \qquad (\text{e.g., m}^3\ \text{s}^{-1}) \quad (4.38)$$

$$= Q/p_s \quad (4.38a)$$

where $V$ is the volume of gas flowing per unit time and $p_s$ is the steady pressure measured at the plane across which $Q$ passes.

Equations 4.38 and 4.38a hold true *if, and only if*, the variation of $p_s$ is negligible over a very long period of time. These equations do not apply to the transient state of flow.

The conversion factors for pumping speed (volume rate of flow) in different systems of units are listed in Table 4.4.

### 4.12.2 Pumping Speed of an Orifice

$$S = Av \qquad (\text{e.g., liters s}^{-1}) \tag{4.39}$$

where $A$ is the area of the orifice and $v$ is the gas velocity (see Eq. 4.12).

When the pressure $p_s$ is sufficiently low, the pumping speed must be calculated for each gas species in the flow.

Even though the conductance ($C$) and the pumping speed ($S$) are expressed using the same units (e.g., $m^3\ h^{-1}$), they have different meanings. Thus, the conductance of a vacuum component is defined as the ratio of throughput under steady-state conservative conditions to the pressure differential between two specified cross sections. The pumping speed is the ratio of the throughput of a given gas to the partial pressure of that gas at a specified point near the inlet port of the pump.

**Table 4.5**

Volume Rate of Flow of an "Ideal Pump" per Unit Port Area ($cm^2$) for Various Species of Gas at Different Temperatures Calculated from Eq. 1.42b

| | liters $s^{-1}$ $cm^{-2}$ | | |
|---|---|---|---|
| Gas | $T = -173°C$ | $T = -73°C$ | $T = 23°C$ |
| Xe | 3.17 | 4.49 | 5.46 |
| Kr | 3.97 | 5.62 | 6.84 |
| $CO_2$ | 5.48 | 7.76 | 9.43 |
| Ar | 5.76 | 8.14 | 9.98 |
| $O_2$ | 6.43 | 9.09 | 11.07 |
| Air | 6.76 | 9.56 | 11.63 |
| $N_2$ | 6.87 | 9.72 | 11.83 |
| CO | 6.87 | 9.72 | 11.83 |
| Ne | 8.09 | 11.45 | 13.93 |
| He | 18.19 | 25.72 | 31.30 |
| $H_2$ | 25.29 | 36.20 | 44.04 |

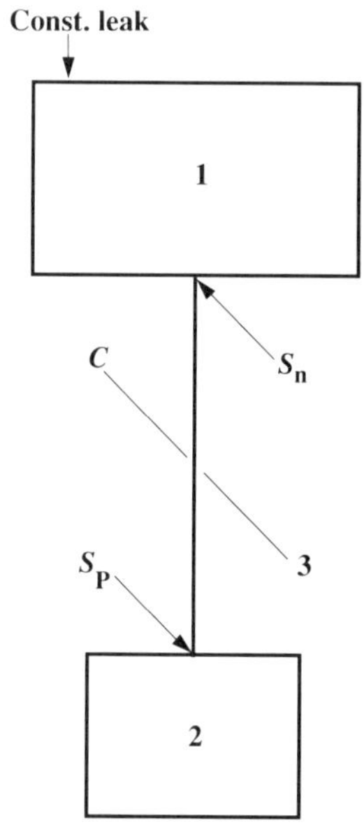


**Figure 4.2** Schematic of a vacuum system: 1—vacuum vessel, 2—vacuum pump, 3—pipe + vacuum components, $S_n$—net pumping speed at the port of the vessel (1), C—conductance of the pipe and of the vacuum components (3), $S_P$—pumping speed at the port of the port of the pump (2).

The volume rates of flow of an "ideal" pump (a pump that evacuates all the molecules impinging on its port of area $A$) calculated from Eq. 3.17b for various species of gas are listed in Table 4.5.

### 4.12.3 Pumping Speed of an Assembly Vacuum Pump—Vacuum Component

Refer to Fig. 4.2.

$$1/S_n = 1/S_P + Z \qquad (\text{e.g., h m}^{-3}) \tag{4.40}$$

$$= 1/S_P + 1/C \tag{4.40a}$$

$$S_n = S_P C/(S_P + C) \qquad (\text{e.g., liters s}^{-1}) \tag{4.41}$$

$$S_P = S_n C/(C - S_n) \qquad (\text{e.g., liters s}^{-1}) \tag{4.42}$$

$$C = S_P S_n/(S_P - S_n) \qquad (\text{e.g., liters s}^{-1}) \tag{4.43}$$

where $Z$ is the impedance and ($C$ the conductance) of the pipe [3 (in Fig. 4.2)]

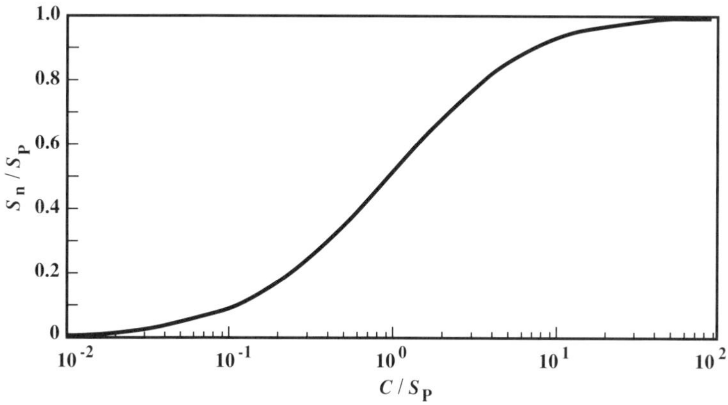


**Figure 4.3** Graph for the evaluation of the net pumping speed $S_n$ at the port of a vessel. [After Delafosse and Mongodin (1961).]

that connects the vacuum pump (2) to the vessel (1) to be evacuated, $S_n$ is the net (effective) pumping speed measured at the port of the vessel (1), and $S_P$ is the pumping speed measured at the port of the pump (2).

Equations 4.40–4.43 hold true only in a permanent regime of flow, i.e., conservative throughput (absence of gas load).

The graph in Fig. 4.3 can be used to evaluate the net pumping speed $S_n$ at the port of a vessel (refer to Fig. 4.2) when the pumping speed $S_P$ at the port of the pump and the conductance C of the pipe connecting the pump to the vessel are known. The graph also permits evaluation of $S_P$ when $S_n$ and C are known, or of C when $S_P$ and $S_n$ are known.

## 4.13 *Basic Considerations in the Design of a Vacuum System*

Keep the pumping speed loss through the pipes of a vacuum system at maximum 20%.

At different sections $1, 2, \ldots, n$ of a vacuum system

$$S_1 p_1 = S_2 p_2 = \cdots S_n p_n = Q \qquad (\text{e.g., mbar liters s}^{-1}) \qquad (4.44)$$

For the design and analysis of a vacuum system, see Santeler (1987).

## *Exercises*

### Exercise 4.1

A quantity of $10^{-3}$ mbar of argon was measured at $T = 23°C$. Calculate (a) the amount of moles and (b) the mass of that amount.

**Solution 4.1**

(a) Use Eq. 4.1b

$$a_{\text{mol}} = a_{\text{pv}}/RT$$

and substitute

$a_{\text{pv}} = 10^{-3}$ mbar at $T = 23°\text{C}$
$R = 83.14$ mbar liters mol$^{-1}$ K$^{-1}$ (see Table 1.3)
$T = 273 + 23 = 296$ K

Then

$$\begin{aligned} a_{\text{mol}} &= 10^{-3} \text{ mbar liters}/83.14 \text{ mbar liters mol}^{-1} \text{ K}^{-1} \times 296 \text{ K} \\ &= 4.06 \times 10^{-8} \text{ mol} \end{aligned}$$

(b) Use Eq. 4.1f

$$a_{\text{kg}} = a_{\text{pv}} M/RT$$

and substitute

$a_{\text{pv}}/RT = a_{\text{mol}} = 4.06 \times 10^{-8}$ mol (see Eq. 4.1a)
$M = 39.948 \times 10^{-3}$ kg mol$^{-1}$ (see Table 1.2)

Thus

$$a_{\text{kg}} = 4.06 \times 10^{-8} \text{ mol} \times 39.948 \times 10^{-3} \text{ kg mol}^{-1} = 1.62 \times 10^{-9} \text{ kg}$$

**Exercise 4.2**

How many molecules are there in a liter of gas at STP?

**Solution 4.2**

According to Eq. 4.1d

$$a_{\text{m}} = a_{\text{pv}}/kT$$

and with

$p = 1013$ mbar (see Section 1.9 on standard conditions for gases at STP)
$V = 1$ liter
$k = 1.381 \times 10^{-22}$ mbar liters molec.$^{-1}$ K$^{-1}$ (see Table 1.3)
$T = 273$ K

there results

$$\begin{aligned} a_{\text{m}} &= 1013 \text{ mbar} \times 1 \text{ liter}/1.381 \times 10^{-22} \text{ mbar liters molec.}^{-1} \text{ K}^{-1} \\ &\quad \times 273 \text{ K} \\ &= 2.69 \times 10^{22} \text{ molec. (per liter)} \end{aligned}$$

*Check*: Calculate the number density $n$ of molecules from Eq. 1.8d

$$n = 7.243 \times 10^{21}\ p/T \qquad (\text{molec. dm}^{-3})$$

and substitute $p = 1013$ mbar, $T = 273$ K. Thus

$$n = 7.343 \times 10^{21} \times 1013/273 = 2.69 \times 10^{22}\ \text{molec. dm}^{-3}$$

The number of molecules in a certain volume does not depend on the gas species.

**Exercise 4.3**

The flow rate of (a) helium, (b) air, and (c) argon gases was measured at $T = 0°C$ and was found to be 1 mbar liter $s^{-1}$. Calculate the mass flow rate of these gases.

**Solution 4.3**

(i) According to Eq. 1.12a written per unit time $t$

$$pV/t = (W_t/t)(RT/M) \qquad \text{at temperature } T$$

(ii) From Eq. 4.1 written per unit time $t$

$$a_{pv}/t = pV/t \quad \text{at temperature } T$$

(iii) Solve for $W_t/M$ the system of equations (i), (ii):

(iv) $$W_t/t = (a_{pv}/t)(M/RT)$$

Substitute in (iv)

$a_{pv}/t = 1$ mbar liter $s^{-1}$
$R = 83.14$ mbar liters $mol^{-1}\ K^{-1}$ (see Table 1.3)
$T = 0°C = 273$ K
$M_{He} = 4.0 \times 10^{-3}$ kg $mol^{-1}$ (see Table 1.2)
$M_{air} = 28.98 \times 10^{-3}$ kg $mol^{-1}$ (see Table 1.2)
$M_{Ar} = 39.9 \times 10^{-3}$ kg $mol^{-1}$ (see Table 1.2)

Thus

(a) For He:

$$W_{t,\text{He}}/t = 1\ \text{mbar liter s}^{-1} \times 4.0 \times 10^{-3}\ \text{kg mol}^{-1}/$$
$$(83.14\ \text{mbar liters mol}^{-1}\ \text{K}^{-1} \times 273\ \text{K}) = 1.76 \times 10^{-4}\ \text{g s}^{-1}$$

(b) For air:

$$W_{t,\text{air}}/t = 1\ \text{mbar liter s}^{-1} \times 28.98 \times 10^{-3}\ \text{kg mol}^{-1}/$$
$$(83.14\ \text{mbar liters mol}^{-1}\ \text{K}^{-1} \times 273\ \text{K}) = 1.28 \times 10^{-3}\ \text{g s}^{-1}$$

(c) For Ar:

$$W_{t,Ar}/t = 1 \text{ mbar liter s}^{-1} \times 38.9 \times 10^{-3} \text{ kg mol}^{-1}/$$
$$(83.14 \text{ mbar liters mol}^{-1}\text{ K}^{-1} \times 273 \text{ K}) = 1.76 \times 10^{-3} \text{ g s}^{-1}$$

**Exercise 4.4**

A 0.05-liter flask filled with gas contains a porous section through which the gas leaks. The pressure rate decrease due to the leak is $1.5 \times 10^{-6}$ mbar liters min$^{-1}$. Calculate the molar flow rate of gas through the porous section, assuming that the temperature of gas that leaks is 23°C.

**Solution 4.4**

Since the volume of the flask is constant, Eq. 4.2 written as

$$q_{mol} = (V/RT)(\Delta p/\Delta t)$$

applies.

Substitute

$$V = 5.0 \times 10^{-2} \text{ liters}$$
$$R = 83.14 \text{ mbar liters mol}^{-1}\text{ K}^{-1} \text{ (see Table 1.3)}$$
$$T = 273 + 23 = 296 \text{ K}$$
$$\Delta p/\Delta t = 1.5 \times 10^{-6} \text{ mbar min}^{-1}$$

Thus

$$q_{mol} = (5.0 \times 10^{-2} \text{ liters}/83.14 \text{ mbar liters mol}^{-1}\text{ K}^{-1} \times 296 \text{ K})$$
$$\times 1.5 \times 10^{-6} \text{ mbar min}^{-1} = 3.0 \times 10^{-12} \text{ mol min}^{-1}$$

**Exercise 4.5**

Vacuum pressure standards of the orifice flow type require flows of gas of $q = 10^{-6}$ mol s$^{-1}$. Express this flow in molecules per second (molec. s$^{-1}$) and grams per second (g s$^{-1}$) for nitrogen.

**Solution 4.5**

Use Eqs. 4.4 and 4.5

$$q_m = q_{mol} N_A$$
$$q_{kg} = q_{mol} M$$

and substitute

$$N_A = 6.022 \times 10^{23} \text{ molec. mol} \quad \text{(see Section 1.10, on Avogadro's number)}$$

$$M_{N_2} = 28 \times 10^{-3} \text{ kg mol}^{-1} \quad \text{(see Table 1.2)}$$

There results

$$q_{\mathrm{m}} = 10^{-6}\ \mathrm{mol\ s^{-1}} \times 6.022 \times 10^{23}\ \mathrm{molec.\ mol^{-1}} = 6.022\ \mathrm{molec.\ s^{-1}}$$

$$q_{\mathrm{kg}} = 10^{-6}\ \mathrm{mol\ s^{-1}} \times 28 \times 10^{-3}\ \mathrm{kg\ mol^{-1}} = 2.8 \times 10^{-8}\ \mathrm{kg\ s^{-1}}$$

**Exercise 4.6**

Hydrogen contained in a volume of 20 cm$^3$ at a pressure of 500 mbar and temperature $T = 23°\mathrm{C}$ escapes through a hole in another volume of 120 cm$^3$ and at a pressure of 70 mbar and temperature $T = 150°\mathrm{C}$. Calculate the average molar flow rate during 10 s.

**Solution 4.6**

Use Eq. 4.3a

$$q_{\mathrm{mol}} = (p_1V_1/RT_1 - p_2V_2/RT_2)/\Delta t$$

and substitute

$p_1 = 500$ mbar
$p_2 = 70$ mbar
$V_1 = 20\ \mathrm{cm}^3 = 0.02$ liters
$V_2 = 120\ \mathrm{cm}^3 = 0.12$ liters
$T_1 = 273 + 23 = 296$ K
$T_2 = 273 + 150 = 423$ K
$R = 83.14$ mbar liters mol$^{-1}$ K$^{-1}$ (see Table 1.3)
$\Delta t = 10$ s

Thus

$$q_{\mathrm{mol}} = [(500\ \mathrm{mbar} \times 0.02\ \mathrm{liters})/(83.14\ \mathrm{mbar\ liters\ mol^{-1}\ K^{-1}} \times 296\ \mathrm{K}) - (70\ \mathrm{mbar} \times 0.12\ \mathrm{liters})/(83.14\ \mathrm{mbar\ liters\ mol^{-1}\ K^{-1}} \times 423\ \mathrm{K})]/10\ \mathrm{s} = 1.67 \times 10^{-5}\ \mathrm{mol\ s^{-1}}$$

**Exercise 4.7**

A gas flow rate of 5 liters s$^{-1}$ at $T = 23°\mathrm{C}$ was measured between the planes $A$ and $B$ perpendicular on the geometric axis of a cylindrical pipe.

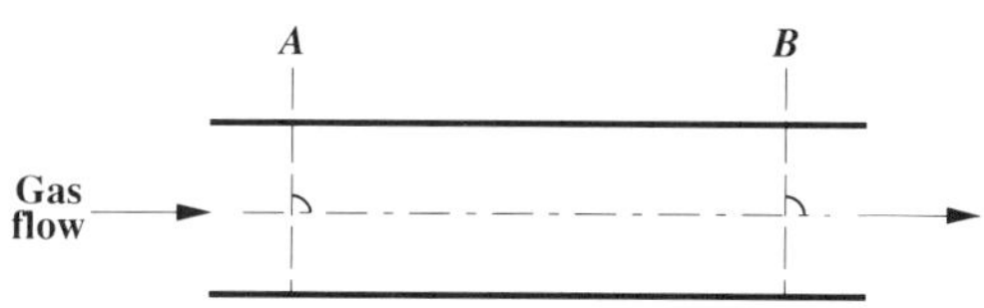


Figure 4A

Calculate the throughput of gas across the plane $B$, assuming that the pressure measured at that plane is 80 mbar. (Refer to Fig. 4A.)

**Solution 4.7**

Use Eq. 4.6a written as

$$Q = p\,\Delta V/\Delta t \qquad (\text{at } T)$$

substitute

$$\begin{aligned} \Delta V/\Delta t &= 5 \text{ liters s}^{-1} \text{ at } T = 23^\circ\text{C} \\ p &= 80 \text{ mbar} \end{aligned}$$

Thus

$$\begin{aligned} Q &= 80 \text{ mbar} \times 5 \text{ liters s}^{-1} = 400 \text{ mbar liters s}^{-1}\ (\text{at } T = 23^\circ\text{C}) \\ &= 40 \text{ W} \quad (\text{at } T = 23^\circ\text{C}) \qquad \left[\text{see Table 4.1 for the conversion of}\right. \\ &\qquad\qquad \left.\text{mbar liters s}^{-1} \text{ to W (watts)}\right] \end{aligned}$$

**Exercise 4.8**

A throughput of argon measured at $T = 0^\circ$C was 0.5 mbar liters s$^{-1}$. Calculate (a) the value of the throughput at $T = 23^\circ$C and (b) the mass flow rates corresponding to the throughputs at $T = 0^\circ$C and $T = 23^\circ$C.

**Solution 4.8**

(a) Use Eq. 4.7 written as

$$Q_2 \quad (\text{at } T_2) = Q_1 \quad (\text{at } T_1) \times (T_2/T_1)$$

Substitute

$$\begin{aligned} Q_1 &= 0.5 \text{ mbar liters s}^{-1}\ (\text{at } T = 0^\circ\text{C}) \\ T_1 &= 273 + 0 = 273 \text{ K} \\ T_2 &= 273 + 23 = 296 \text{ K} \end{aligned}$$

Then

$$\begin{aligned} Q_2 &= 0.5 \text{ mbar liters s}^{-1} \times 296 \text{ K}/273 \text{ K} \\ &= 0.542 \text{ mbar liters s}^{-1} \quad (\text{at } T = 23^\circ\text{C}) \end{aligned}$$

The error resulting from considering the throughput measured at $T = 23^\circ$C instead of that measured at $T = 0^\circ$C is

$$100(0.542 \text{ mbar liters s}^{-1} - 0.5 \text{ mbar liters s}^{-1})/0.5 \text{ mbar liters s}^{-1} = 8.4\%$$

(b) From Eq. 4.10

$$q_{kg} = Q_{Mg\,Tg} M_g / RT$$

Substitute

$M_g = 39.94 \times 10^{-3}$ kg mol$^{-1}$ (see Table 1.2)
$R = 83.14$ mbar mol$^{-1}$ K$^{-1}$ (see Table 1.3)

(i) $Q_{Mg\,Tg} = 0.5$ mbar liters s$^{-1}$ (at $T = 0°C$)
(ii) $Q_{Mg\,Tg} = 0.542$ mbar liters s$^{-1}$ (at $T = 23°C$)
(i) $T_g = 273 + 0 = 273$ K
(ii) $T_g = 273 + 23 = 296$ K

Thus

(i) $q_{kg} = 0.5$ mbar liters s$^{-1}$ $\times$ 39.9 $\times$ $10^{-3}$ kg mol$^{-1}$/ (83.14 mbar liters mol$^{-1}$ K$^{-1}$ $\times$ 273 K)
$= 878.96 \times 10^{-9}$ kg s$^{-1}$ (at $T = 0°C$)

(ii) $q_{kg} = 0.542$ mbar liters s$^{-1}$ $\times$ 39.9 $\times$ $10^{-3}$ kg mol$^{-1}$/ (83.14 mbar liters mol$^{-1}$ K$^{-1}$ $\times$ 296 K)
$= 878.76 \times 10^{-9}$ kg s$^{-1}$ (at $T = 23°C$)

**Exercise 4.9**

Calculate the throughput in atm cm$^3$ s$^{-1}$ at $T = 23°C$, corresponding to a helium flow rate of $8.5 \times 10^{-11}$ mol s$^{-1}$ at $T = 23°C$.

**Solution 4.9**

(i) Convert the molar flow rate of helium at $T = 23°C$ to a mass flow rate at the same temperature. Use Eq. 4.5:

$$q_{kg} = q_{mol} M$$

Substitute

$q_{mol} = 8.5 \times 10^{-11}$ mol s$^{-1}$ (at $T = 23°C$)
$M = 4.0 \times 10^{-3}$ kg mol$^{-1}$ (see Table 1.2)

Then

$$k_{kg} = 8.5 \times 10^{-11} \text{ mol s}^{-1} \times 4.0 \times 10^{-3} \text{ kg mol}^{-1}$$

$$= 3.4 \times 10^{-13} \text{ kg s}^{-1} \quad (\text{at } T = 23°C)$$

(ii) Convert the helium mass flow rate calculated at (i) to throughput of helium at $T = 23°C$ using Eq. 4.10 written as

$$Q \ (\text{at } T = 23°C) = q_{kg} \ (\text{at } T = 23°C) \times RT/M$$

Substituting

$$q_{kg}\text{ (at } T = 23°\text{C)} = 3.4 \times 10^{-13}\text{ kg s}^{-1}$$
$$R = 82.058\text{ atm cm}^3\text{ mol}^{-1}\text{ K}^{-1}\text{ (see Table 1.3)}$$
$$T = 273 + 23 = 296\text{ K}$$

There results

$$Q = 3.4 \times 10^{-13} \times 82.058\text{ atm cm}^3\text{ mol}^{-1}\text{ K}^{-1} \times 296\text{ K}/$$
$$4.0 \times 10^{-3}\text{ kg kmol}^{-1} = 2.06 \times 10^{-9}\text{ atm cm}^3\text{ s}^{-1} \quad \text{(at } T = 23°\text{C)}$$

**Exercise 4.10**

The throughput of a rotary-vane vacuum pump in the viscous range for nitrogen is 380 mbar $m^3$ $h^{-1}$ at $T = 15°C$. The throughput of a diffusion pump in the molecular range for nitrogen is $10^{-3}$ mbar liters $s^{-1}$ at $T = 15°C$. Calculate the throughput of these pumps for hydrogen at $T = 15°C$.

**Solution 4.10**

To convert throughputs from one species of gas to another for constant temperature conditions, use Eqs. 4.8 and 4.9:

$$Q_{l1} = Q_{l2}(\eta_2/\eta_1) \qquad \text{(laminar flow)}$$

$$Q_{m1} = Q_{m2}(M_2/M_1)^{1/2} \qquad \text{(molecular flow)}$$

(i) Substitute in the equation for the conversion of throughputs in the laminar flow range (subscript l)

$$Q_2 = 380\text{ mbar m}^3\text{ h}^{-1} \qquad (N_2 \text{ at } T = 15°\text{C})$$
$$\eta_2 = 170.8\ \mu\text{poise} \qquad (N_2 \text{ at } T = 15°\text{C; see Table 3.12})$$
$$\eta_1 = 91.0\ \mu\text{poise} \qquad (H_2 \text{ at } T = 15°\text{C; see Table 3.12})$$

Thus

$$Q_{l1} = 380\text{ mbar m}^3\text{ h}^{-1} \times (170.8\ \mu\text{poise}/91.0\ \mu\text{poise})$$
$$= 713\text{ mbar m}^3\text{ h}^{-1} \quad (H_2 \text{ at } T = 15°\text{C})$$

(ii) Substitute in the equation for the conversion of throughputs in the molecular flow range (subscript m)

$$Q_{m2} = 10^{-3}\text{ mbar liters s}^{-1} \qquad (N_2 \text{ at } T = 15°\text{C})$$

$$M_{2,N_2} = 28 \times 10^{-3}\text{ kg mol}^{-1} \qquad \text{(see Table 1.2)}$$

$$M_{1,H_2} = 2 \times 10^{-3}\text{ kg mol}^{-1} \qquad \text{(see Table 1.2)}$$

Then

$$Q_{ml} = 10^{-3} \text{ mbar liters s}^{-1}$$

$$\times (28 \times 10^{-3} \text{ kg mol}^{-1}/2 \times 10^{-3} \text{ kg mol}^{-1})^{1/2}$$

$$= 3.74 \times 10^{-3} \text{ mbar liters s}^{-1} \quad (H_2 \text{ at } T = 15°\text{C})$$

**Exercise 4.11**

Prove that the Reynolds number Re is dimensionless.

**Solution 4.11**

From Eq. 4.11

$$\text{Re} = dv\,\rho/\eta$$

Using the SI system, substitute

$d$ (m)
$v$ (m s$^{-1}$)
$\rho$ (kg m$^{-3}$)
$\eta$ (Pa s or kg m$^{-1}$ s$^{-1}$)

Then

$$\text{Re} = (\text{m}) \times (\text{m s}^{-1}) \times (\text{kg m}^{3})/(\text{kg m}^{-1}\text{ s}^{-1}) \text{ dimensionless}$$

Using the CGS system, substitute

$d$ (cm)
$v$ (cm s$^{-1}$)
$\rho$ (g cm$^{-3}$)
$\eta$ (pose or g cm$^{-1}$ s$^{-1}$)

Thus

$$\text{Re} = (\text{cm}) \times (\text{cm s}^{-1}) \times (\text{g cm}^{-3})/(\text{g cm}^{-1}\text{ s}^{-1}) \text{ dimensionless}$$

**Exercise 4.12**

Infer the throughput–Reynolds number relationship that characterizes the viscous flow of air in a pipe of circular cross section.

**Solution 4.12**

(i) Use Eqs. 4.11, 4.12, and 1.14b

$$\text{Re} = dv\,\rho/\eta$$

$$v = V'/(\pi d^2/4)$$

$$\rho = pm/kT$$

and solve for Re

(ii) $$\mathrm{Re} = 4V'pm/(\pi\, dkT\eta)$$

Substitute $V'$ by $Q/p$ (Eq. 4.6b) and solve for $Q$

$$Q = (\mathrm{Re}\, d)(\pi kT\eta/4m)$$

which is Eq. 4.15a.

**Exercise 4.13**

Air at $T = 23°\mathrm{C}$ flows through a pipe due to a pressure difference of 500 mbar at the ends of the pipe. The measured mass flow rate of air is 20 kg s$^{-1}$. Calculate (a) the impedance of the pipe to the air flow, (b) the throughput of air, and (c) the conductance of the pipe for air.

**Solution 4.13**

(a) Use Eq. 4.30:

$$Z = (1/q_{\mathrm{kg}})(p_1 - p_2)M/RT$$

Substitute

$q_{\mathrm{kg}} = 20$ kg s$^{-1}$ (at $T = 23°\mathrm{C}$)
$p_1 - p_2 = 500$ mbar
$M = 28.98 \times 10^{-3}$ kg mol$^{-1}$ (see Table 1.2)
$R = 83.14$ mbar liters mol$^{-1}$ K$^{-1}$ (see Table 1.3)
$T = 273 + 23 = 296$ K

Thus

$$Z = (1/20\ \mathrm{kg\ s^{-1}}) \times 500\ \mathrm{mbar} \times 28.98 \times 10^{-3}\ \mathrm{kg\ mol^{-1}}/$$
$$83.14\ \mathrm{mbar\ liters\ mol^{-1}\ K^{-1}} \times 296\ \mathrm{K}$$
$$= 2.94 \times 10^{-5}\ \mathrm{s\ liter^{-1}}\ (\mathrm{at}\ T = 23°\mathrm{C})$$

(b) From Eq. 4.6c:

$$Q\ (\mathrm{at}\ T = 23°\mathrm{C}) = q_{\mathrm{kg}}\ (\mathrm{at}\ T = 23°\mathrm{C})RT/M$$

$$Q = 20\ \mathrm{kg\ s^{-1}} \times 83.13\ \mathrm{mbar\ liters\ mol^{-1}\ s^{-1}} \times 296\ \mathrm{K}/$$
$$28.98 \times 10^{-3}\ \mathrm{kg\ mol^{-1}}$$
$$= 1.7 \times 10^{7}\ \mathrm{mbar\ liters\ s^{-1}}\quad (\mathrm{at}\ T = 23°\mathrm{C})$$

(c) Approach 1:
Use Eq. 4.31:

$$C = 1/Z$$

Substitute

$$Z = 2.94 \times 10^{-5} \text{ s liter}^{-1} \quad (\text{at } T = 23°\text{C})$$

Then

$$C = 1/2.94 \times 10^{-5} \text{ s liter}^{-1} = 3.4 \times 10^{4} \text{ liters s}^{-1} \quad (\text{at } T = 23°\text{C})$$

(d) Approach 2:
Use Eq. 4.31b:

$$C = Q/(p_1 - p_2)$$

Substitute

$$Q = 1.7 \times 10^{7} \text{ mbar liters s}^{-1} \text{ (see Eq. 4.31b)}$$

$$p_1 - p_2 = 500 \text{ mbar}$$

Then

$$C = 1.7 \times 10^{7} \text{ mbar liters s}^{-1}/500 \text{ mbar}$$

$$= 3.4 \times 10^{4} \text{ s}^{-1}/500 \text{ mbar} = 3.4 \times 10^{4} \text{ s}^{-1} \text{ (at } T = 23°\text{C)}$$

**Exercise 4.14**

Consider two pipes connected in series. Determine the total conductance of the assembly, assuming that the conductance of the individual pipes are (a) 150 liters $s^{-1}$ and 120 liters $s^{-1}$, (b) 150 liters $s^{-1}$, and 50 liters $s^{-1}$. Comment on the results.

**Solution 4.14**

Use Eq. 4.33a for both cases (a) and (b):

$$C_{1,2} = C_1C_2/(C_1 + C_2)$$

(a) Substitute

$$C_1 = 150 \text{ liters s}^{-1}$$

$$C_2 = 120 \text{ liters s}^{-1}$$

Thus

$$C_{1,2} = 150 \text{ liters s}^{-1} \times 120 \text{ liters s}^{-1}/(150 \text{ liters s}^{-1} + 120 \text{ liters s}^{-1})$$

$$= 66.7 \text{ liters s}^{-1}$$

(b) Substitute

$$C_1 = 150 \text{ liters s}^{-1}$$

$$C_2 = 50 \text{ liters s}^{-1}$$

Thus

$$C_{1,2} = 150 \text{ liters s}^{-1} \times 50 \text{ liters s}^{-1}/(150 \text{ liters s}^{-1} + 50 \text{ liters s}^{-1})$$
$$= 37.5 \text{ liters s}^{-1}$$

*Comments*: If the conductance of the pipes connected in series $C_{1,2}$ is reduced to the conductance of the pipe $C_1$, for both cases (a) and (b), the following results:

(a) If $C_{1,2}/C_1 = 66.7$ liters $s^{-1}/150$ liters $s^{-1}$, 44.6% of the conductance $C_1$ is used and 100% − 44.6% = 55.4% is lost.
(b) If $C_{1,2}/C_1 = 37.5$ liters $s^{-1}/150$ liters $s^{-1}$, 25% of the conductance $C_1$ is used and 100% − 25% = 75% is lost.

Case (b) is an example of inefficient connection of pipes in series.

**Exercise 4.15**
Two pipes of conductance 500 liters $s^{-1}$ and 400 liters $s^{-1}$, respectively, are connected in series. Use the nomogram in Fig. 4.1 to evaluate the conductance of the assembly.

**Solution 4.15**

(a) Use the same units (in this case liters $s^{-1}$) for all the scales of the nomograph.
(b) Multiply by 10 all the scales of the nomograph.
(c) The straight line that joins the value 50 liters $s^{-1}$ (actually 500 liters $s^{-1}$) on the scale $C_1$ with the value 40 liters $s^{-1}$ (actually 400 liters $s^{-1}$) on the scale $C_2$ intersects the scale $C_3$ at 22.2 liters $s^{-1}$ (actually 222 liters $s^{-1}$). The conductance of the assembly is therefore 222 liters $s^{-1}$.

**Exercise 4.16**
Two vacuum components, each one having a conductance of 450 liters $s^{-1}$ joined in parallel, are connected in series to a pipe having a conductance of 500 liters $s^{-1}$. What is the conductance of the assembly?

**Solution 4.16**

(i) To calculate the total conductance of the vacuum components joined in parallel, use Eq. 4.35a:

$$C_{1,2} = C_1 + C_2$$

Substitute

$$C_1 = C_2 = 450 \text{ liters s}^{-1}$$

Thus

$$C_{1,2} = 450 \text{ liters s}^{-1} + 450 \text{ liters s}^{-1} = 900 \text{ liters s}^{-1}$$

(ii) To calculate the conductance of the assembly of vacuum components connected in parallel to the pipe, use Eq. 4.33a written as

$$C_{\text{assembly}} = C_{1,2}C_3/(C_{1,2} + C_3)$$

Substitute

$$C_{1,2} = 900 \text{ liters s}^{-1}$$

$$C_3 = 500 \text{ liters s}^{-1}$$

Thus

$$C_{\text{assembly}} = 900 \text{ liters s}^{-1} \times 500 \text{ liters s}^{-1}/(900 \text{ liters s}^{-1} + 500 \text{ liters s}^{-1})$$

$$= 321.4 \text{ liters s}^{-1}$$

**Exercise 4.17**

Consider the schematic of the vacuum system illustrated in Fig. 4.2 and suppose that a vacuum pump of known pumping speed is available. Use the graph in Fig. 4.3 to determine the conductance of the line connecting the pump to the vessel such as to obtain a maximum pumping speed at the port of the vessel. Comment on the results.

**Solution 4.17**

Let us assume that

(i) $$C = 0.1S_P \quad \text{or} \quad C/S_P = 0.1$$

For this value on the abscissa (horizontal, $x$ axis) of the graph, the value corresponding to the ordinate (vertical, $y$) axis is

$$S_n/S_P = 0.09, \quad \text{i.e.,} \quad S_n = 9\%S_P$$

(ii) $$C = S_P \quad \text{or} \quad C/S_P = 1$$

The corresponding value on the ordinate axis is

$$S_n/S_P = 0.5 \quad \text{or} \quad S_n = 50\%S_p$$

(iii) $$C = 10S_P \quad \text{or} \quad C/S_P = 10$$

Again the corresponding value on the ordinate axis is

$$S_n/S_P = 0.905 \quad \text{or} \quad S_n = 90.5\%S_P$$

*Comments*: The net pumping speeds $S_n$ at the port of the vessel (1 in Fig. 4.2) are as follows:

(i) $S_n = 0.09 S_P$, i.e., $100\% S_P - 9\% S_P = 99.91\%$ loss through the conductance C.
(ii) $S_n = 0.5 S_P$, i.e., $100\% S_P - 50\% S_P = 50\% S_P$ loss through the conductance C.
(iii) $S_n = 0.905 S_P$, i.e., $100\% S_P - 90.5\% S_P = 9.5\%$ loss through the conductance C.

Thus, the factor that limits the pumping speed $S_n$ at the port of the vessel (1) is the conductance of the line (3) that connects the pump (2) to the vessel. An increase in the pumping speed $S_P$ of the pump would not be useful in obtaining a higher value $S_n$ at the port of the vessel.

**Exercise 4.18**

What pump is required to maintain a pumping speed of $10^{-7}$ mbar in a vessel where air at $T = 23°\text{C}$ leaks at a rate of 2 cm$^3$ min$^{-1}$ and at a pressure of $10^{-1}$ mbar?

**Solution 4.18**

Use Eqs. 4.38a and 4.6b written as

$$S_n = Q/p_s$$

$$Q_{\text{leak}} \quad (\text{at } T) = p_{\text{leak}} V'$$

Substitute

$$\begin{aligned} p_{\text{leak}} &= 10^{-1} \text{ mbar} \\ V' &= 2 \text{ cm}^3 \text{ min}^{-1} = 2 \text{ cm}^3 \times (1/100) \text{ liters cm}^{-3} \times (1/60) \text{ s min}^{-1} \\ &= 3.3 \times 10^{-5} \text{ liters s}^{-1} \\ p_s &= 10^{-7} \text{ mbar} \end{aligned}$$

Thus

$$Q_{\text{leak}} = 10^{-1} \text{ mbar} \times 3.3 \times 10^{-5} \text{ liters s}^{-1} = 3.3 \times 10^{-6} \text{ mbar liters s}^{-1}$$

$$S_n = 3.3 \times 10^{-6} \text{ mbar liters s}^{-1} / 10^{-7} \text{ mbar} = 33 \text{ liters s}^{-1}$$

In order to handle this permanent gas load and some additional loads, such as gas bursts from the diffusion pump (if the vessel is evacuated by such a pump), outgassing, as well as losses in the line connecting the vacuum pump to the vessel a safety factor 3, is recommended for the choice of the pump. Thus a vacuum pump having a pumping speed

$$S_p = 33.3 \text{ liters s}^{-1} \times 3 \cong 100 \text{ liters s}^{-1}$$

is necessary.

## References

Delafosse, J., and Mongodin, G. (1961), *Le Vide*, *92*, 107.

Diels, K., and Jaeckel, R. (1958), Leybold Vakuum-Taschenbuch, Springer-Verlag, Berlin.

Knudsen, M. (1910), *Ann. Physik*, *31*, 205; *32*, 890; *38*, 1435.

Lawrence, T., and Lamont, R., Jr. (1979), in *Methods of Experimental Physics*, Vol. 14, *Vacuum Physics and Technology*, G. L. Weissler and R. W. Carlson (eds.), Academic Press, New York.

Moss, T. A. (1987), *J. Vac. Sci. Technol.* A5, 5, 2962.

Santeler, D. J. (1987), *J. Vac. Sci. Technol.* A5, 4, 2472.

Wilson, R. S. (1987), *J. Vac. Sci. Technol.* A5, 4, 2479.

# 5

# Steady Flow of Gas in the Viscous Range

## 5.1 *Turbulent Flow*

At high pressures and velocities, gas flow becomes erratic with the appearance of eddies and oscillations. Such flow, termed *turbulent* (denoted by subscript t), can be predicted by evaluating the Reynolds number (Eq. 4.13) or the throughput–Reynolds number relationship (Eqs. 4.16–4.18).

In vacuum practice, turbulent flow may occur at the initiation of the roughing process, and can be prevented by slowly opening the roughing valve.

## 5.2 *Laminar Flow through an Aperture*

When gas at constant pressure $p_1$ (e.g., atmospheric pressure) contained in a large region escapes through a small aperture into a region where pressure is $p_2$ ($p_2 < p_1$) and can be controlled, the gas acquires a velocity toward the aperture and streamlines are set up in the direction of the flow (Fig. 5.1). The gas reaches a minimum cross section (called *vena contracta*) on the low-pressure side $p_2$, and then expands and contracts again approximately 12 times. Soon afterward the expansion–contraction pattern loses its identity as a result of diffusion and turbulence. This is *laminar* flow (denoted by subscript l).

### 5.2.1 Laminar Nonchocked Throughput

This type of throughput is denoted by subscript lnc:

$$Q_{\mathrm{lnc}} = \left[(2\nu/\nu - 1)(RT_1/M)^{1/2}(p_2/p_1)^{1/\nu}\right]^{1/2} \times\left[1 - (p_2/p_1)^{(\nu-1)/\nu}\right]^{1/2} p_1 A \qquad (\text{ergs s}^{-1}) \qquad (5.1)$$

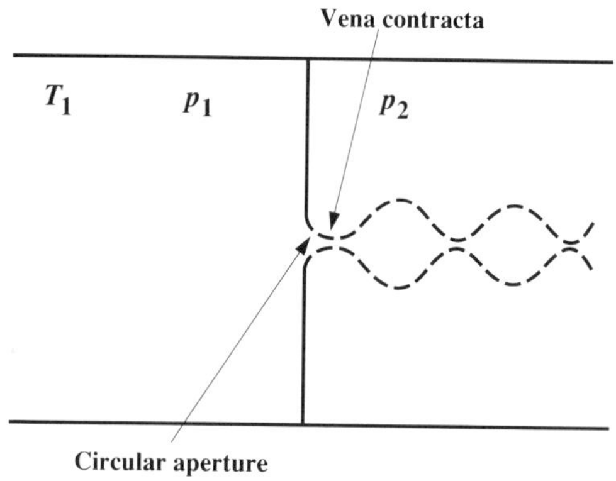


**Figure 5.1** Schematic of gas flow through a "thin" small aperture.

where $\nu$ (dimensionless; see Section 3.13)
$R$ (ergs mol$^{-1}$ K$^{-1}$)
$T_1$ (K)
M (g mol$^{-1}$)
$p_1$, $p_2$ (dyn cm$^{-2}$)
A (cm$^2$)

For air at $T = 23°C$, $\nu = 1.403$ (dimensionless; see Table 3.14), M = 28.98 g mol$^{-1}$ (see Table 1.2), and with $R = 8.314 \times 10^7$ ergs mol$^{-1}$ K$^{-1}$ (see Table 1.3), $p_1$, $p_2$ (dyn cm$^{-2}$), A (cm$^2$), Eq. 5.1 becomes

$$Q_{\text{Inc}} = 76894(p_2/p_1)^{0.713}\left[1 - (p_2/p_1)^{0.287}\right]^{1/2} p_1 A \qquad (\text{dyn cm}^{-2}\ \text{cm}^3\ \text{s}^{-1}) \quad (5.2)$$

Equation 5.2 holds true if the diameter of the aperture is very large compared to the mean free path of gas molecules.

When the velocity of gas molecules crossing the aperture equals the velocity of sound in that gas, $Q_{\text{Inc}}$ reaches its maximum value. This occurs when the Mach number Ma is unity.

The Mach number is defined as

$$\text{Ma} = v_A/v_s \qquad (\text{dimensionless}) \qquad (5.3)$$

where $v_A$ (cm s$^{-1}$) is the average velocity of the gas flow across a certain plane and $v_s$ (cm s$^{-1}$) is the local acoustic velocity, which is a function of the gas species ($\nu$, M) and temperature. Thus

$$v_s = 8700(\nu T/M) \qquad (\text{cm s}^{-1}) \tag{5.4}$$

and for air at $T = 23°\text{C}$

$$v_s = 3.29 \times 10^4 \qquad (\text{cm s}^{-1}) \tag{5.4a}$$

The boundary values of the gas throughput (Eq. 5.1) across the aperture are

$$Q_{\text{lnc}} = 0 \quad \text{for} \quad p_2/p_1 = 0 \tag{5.5}$$

$$Q_{\text{lnc}} = \text{max. for } p_2/p_1 = r_c = \left[2/(\nu + 1)^{\nu/(\nu-1)}\right]$$
$$= \text{critical pressure (dimensionless)} \tag{5.6}$$

For air at $T = 23°\text{C}$ the critical pressure has the value

$$r_c = 0.528 \tag{5.6a}$$

### 5.2.2 Laminar Nonchocked Conductance

This type of conductance is denoted by subscript lnc:

$$C_{\text{lnc}} = \{1/(1 - p_2/p_1)[2\nu/(\nu - 1)](RT_1/M)\}^{1/2}\left[(p_2/p_1)^{1/\nu}\right]^{1/2}$$
$$\times\left[1 - (p_2/p_1)^{(\nu-1)/\nu}\right]^{1/2} p_1 A \qquad (\text{cm}^3\ \text{s}^{-1}) \tag{5.7}$$

For air at $T = 23°\text{C}$

$$C_{\text{lnc}} = [76.894/(1 - p_2)/p_1](p_2/p_1)^{0.713}\left[1 - (p_2/p_1)^{0.287}\right]^{1/2} A$$
$$(\text{liters s}^{-1}) \tag{5.7a}$$

### 5.2.3 Laminar Chocked Conductance

This type of conductance is denoted by subscript lch:

$$C_{\text{lch}} = O_{\text{ch}}\left[(\nu RT_1/M)\left(2/(\nu + 1)^{(\nu+1)/(\nu-1)}\right)\right]^{1/2} A \qquad (\text{cm}^3\ \text{s}^{-1}) \tag{5.8}$$

where $O_{\text{ch}}$ (dimensionless) is an aperture (orifice) coefficient. This coefficient allows for vena contracta (the squeezing of the cross-sectional area of the gas beam in viscous state, which enters the aperture). The value of $O_{\text{ch}}$ varies with both the geometry of the aperture and the pressure, from 1.0 (well-rounded entrance) to 0.75 (Santeler, 1986; Santeler et al., 1966, p. 93).

For air at $T = 23°\text{C}$

$$C_{\text{lch}} = 3.54 O_{\text{ch}} A \qquad (\text{liters s}^{-1}) \tag{5.9}$$

### 5.2.4 Laminar Chocked Throughput

This type of throughput is denoted by subscript lch:

$$Q_{\text{lch}} = C_{\text{lch}}\,p_2 \qquad (\text{e.g., mbar liters s}^{-1}) \tag{5.10}$$

## 5.3 *Laminar Flow through Long Pipes*

A pipe (or channel) is considered *long* when the conductance it offers to gas flow is small compared to the conductance of its entrance aperture for the same gas. Practically, a pipe of length $L$ and diameter $d$ is considered long when $L \geqslant 20d$. This type of flow is denoted by subscript l.

### 5.3.1 Poiseuille's Law

The viscous flow rate of gas through a straight pipe of uniform circular cross section is proportional to the pressure differential at the ends of the pipe and the radius of the pipe at the 4th power and inversely proportional to the length of the pipe and the gas viscosity. Thus

$$Q_{\text{l}} = (\pi r^4/8\eta L)\bar{p}(p_1 - p_2) \qquad (\text{ergs s}^{-1}) \tag{5.11}$$

$$= (\pi d^4/128\eta L)\bar{p}(p_1 - p_2) \tag{5.11a}$$

where
$r$ = radius of the pipe (cm)
$d$ = diameter of the pipe (cm)
$\eta$ = gas viscosity (poise)
$L$ = length of the pipe (cm)
$p_1, p_2$ = pressures at entrance and exit of the pipe, respectively (dyn cm$^{-2}$)
$\bar{p}$ = average pressure $\bar{p} = (p_1 + p_2)/2$ (dyn cm$^{-2}$)

*a. Laminar throughput of straight pipes of uniform circular cross section* For air at $T = 23°\text{C}$ the rate of flow $Q_{\text{l}}$ expressed in different system of units is

$$Q_{\text{l}} = 4.97(d^4/L)\bar{p}(p_1 - p_2) \qquad (\text{Pa m}^3\text{ h}^{-1}) \tag{5.11b}$$

with $d$ (m), $L$ (m), $p_1, p_2$ (Pa),

$$Q_{\text{l}} = 138\,(d^4/L)\bar{p}(p_1 - p_2) \qquad (\text{mbar liters s}^{-1}) \tag{5.11c}$$

with $d$ (cm), $L$ (cm), $p_1, p_2$ (mbar), and

$$Q_{\text{l}} = 184\,(d^4/L)\bar{p}(p_1 - p_2) \qquad (\text{Torr liters s}^{-1}) \tag{5.11d}$$

with $d$ (cm), $L$ (cm), $p_1, p_2$ (Torr).

Equations 5.11–5.11d hold true provided that

1. There is negligible compressibility of gas, i.e., the Mach number is Ma 0.3 (see Eq. 5.3).
2. The profile of the flow velocity is constant throughout the length $L$.
3. There is no turbulent motion of gas.
4. The gas velocity at the wall of the pipe equals zero, i.e., the ratio $\lambda/d$ is very large.

*b. Laminar conductance of straight pipes of uniform circular cross section*

$$C_l = (\pi d^4/128\eta L)\bar{p} \qquad (\text{cm}^3\ \text{s}^{-1}) \tag{5.12}$$

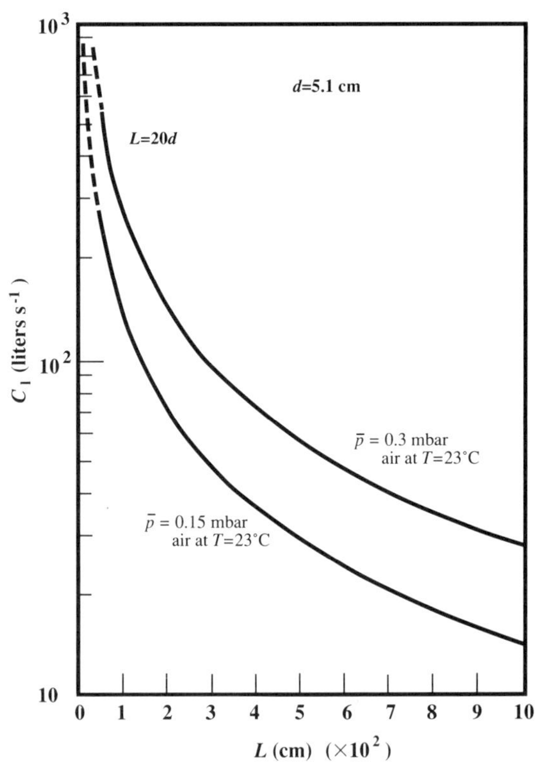


**Figure 5.2** Laminar flow conductance C for air at $T = 23°\text{C}$ of a long straight pipe of uniform circular cross section, $d = 5.1$ cm for various average pressures (mbar), calculated from Eq. 5.12b.

For air at $T = 23°C$ the conductance, expressed in different systems of units, is

$$C_l = 4.97(d^4/L)\bar{p} \qquad (m^3\ h^{-1}) \tag{5.12a}$$

with $d$ (m), $L$ (m), $\bar{p}$ (Pa),

$$C_l = 138(d^4/L)\bar{p} \qquad (\text{liters } s^{-1}) \tag{5.12b}$$

with $d$ (cm), $L$ (cm), $\bar{p}$ (mbar), and

$$C_l = 184(d^4/L)\bar{p} \qquad (\text{liters } s^{-1}) \tag{5.12c}$$

with $d$ (cm), $L$ (cm), $\bar{p}$ (Torr).

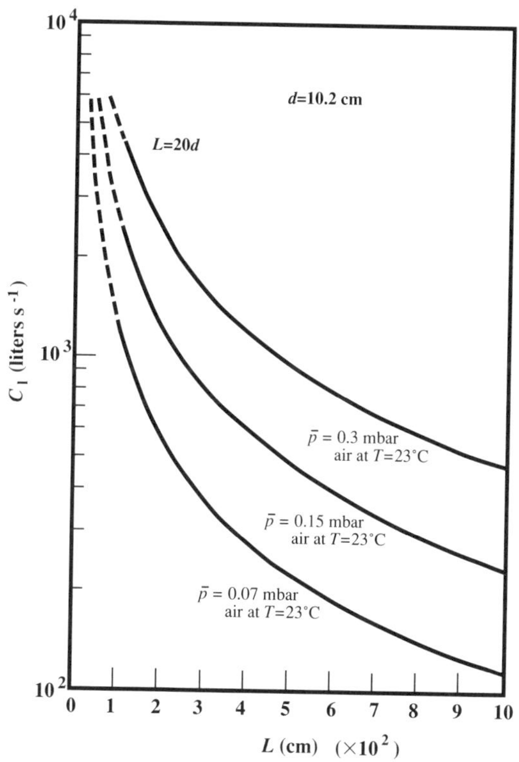


**Figure 5.3** Laminar flow conductance $C_l$ for air at $T = 23°C$ of a long straight pipe of uniform circular cross section, $d = 10.2$ cm for various average pressures (mbar), calculated from Eq. 5.12b.

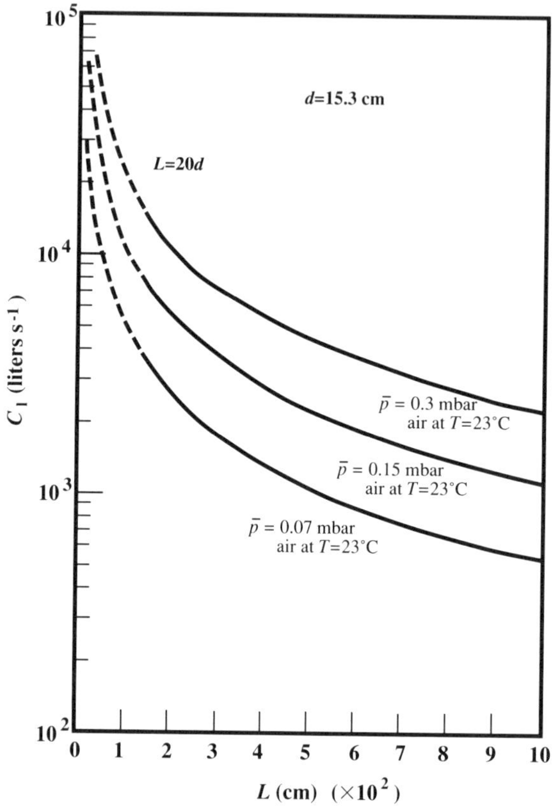


**Figure 5.4** Laminar flow conductance $C_l$ for air at $T = 23°C$ of a long straight pipe of uniform circular cross section, $d = 15.3$ cm for various average pressures (mbar), calculated from Eq. 5.12b.

Equations 5.12a and 5.12b are accurate to 10% for $\bar{p}d \geqslant 6.5 \times 10^{-1}$ when the average pressure is expressed in pascals or millibars and the diameter is expressed in meters or centimeters, i.e., with $\bar{p}$ (Pa or mbar), and $d$ (m or cm). Equation 5.12c is accurate to 10% for $\bar{p}d \geqslant 4.9 \times 10^{-1}$ with $\bar{p}$ (Torr), $d$ (cm) (Guthrie and Wakelring, 1949, p. 30).

The laminar flow conductance of long pipes of circular cross section of various inner diameters is plotted in the graphs in Figs. 5.2–5.5. These graphs help in selecting the length of a pipe as a function of the average pressure $\bar{p}$.

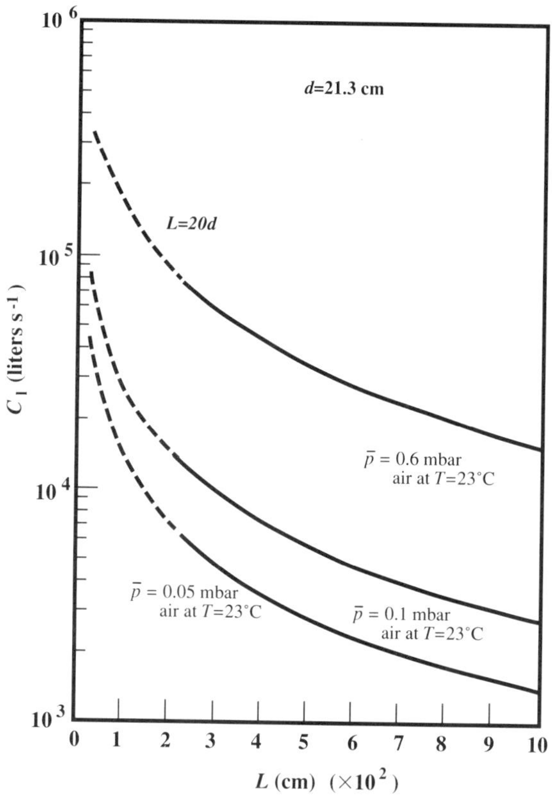


**Figure 5.5** Laminar flow conductance $C_l$ for air at $T = 23°C$ of a long straight pipe of uniform circular cross section, $d = 21.3$ cm for various average pressures (mbar), calculated from Eq. 5.12b.

### 5.3.2 Laminar Volume Flow Rate of Gas through a Pipe

If $V'$ is the volume rate of gas flow at a specified temperature (see Eq. 4.12) at any plane through the pipe, then the following relationships are available for air at $T = 23°C$:

$$V' = 138(d^4/L)(p_1 - p_2) \qquad (\text{liters s}^{-1}) \tag{5.13}$$

with $d$ (cm), $L$ (cm), $p_1$, $p_2$ (mbar) and

$$V' = 184(d^4/L)(p_1 - p_2) \qquad (\text{liters s}^{-1}) \tag{5.14}$$

with $d$ (cm), $L$ (cm), $p_1$, $p_2$ (Torr).

### 5.3.3 Pressure-Drop Formula

For the efficient utilization of the pump capacity, the pressure drop along the pipe that connects the pump to the vessel to be evacuated, must not exceed one-fifth (or 20% of) the pressure at the port of the pump at *operating level.*

A rough check of the size of a pipe of length $L$ and diameter $d$, for air at $T = 23°C$, when pressure is expressed in different systems of units, is

$$p_1 - p_P = 0.20 S_P = 0.20 S_P L/d^4 \qquad \text{(Pa)} \tag{5.15}$$

where
$p_1$ (Pa) = the pressure at the port of the vessel evacuated
$p_P$ (Pa) = the pressure at the port of the pump *at operating level*
$S_P$ ($m^3\ h^{-1}$) = the pumping speed at the port of the pump
$L$ (m) = the length of the pipe
$d$ (m) = the inner diameter of the pipe

and

$$p_1 - p_P = 7.2 \times 10^{-3} S_P L/d^4 \qquad \text{(mbar)} \tag{5.15a}$$

with $p_1$, $p_P$ (mbar), $S_P$ (liters $s^{-1}$), $L$ (cm), $d$ (cm) and

$$p_1 - p_P = 5.43 \times 10^{-3} S_P L/d^4 \qquad \text{(Torr)} \tag{5.15b}$$

with $p_1$, $p_P$ (Torr), $S_P$ (liters $s^{-1}$), $L$ (cm), $d$ (cm).

The value $S_P = f(p_1)$ can be found on the performance characteristic curve of the pump.

If $(p_1 - p_P)$ calculated is large compared to the desired operating pressure $p_P$, a larger pipe size must be selected.

### 5.3.4 Correction to Poiseuille's Law owing to Surface Slip

In the derivation of Poiseuille's law it was assumed that the gas velocity at the wall of the pipe equals zero. However, owing to the gas molecules adsorbed or remitted at the wall, a viscous drag effect causes the velocity of gas to differ from zero. Poiseuille's law (Eq. 5.11), corrected for this effect, known as "surface slip" (Van Atta, 1965, p. 34) is

$$Q = (c_1 \bar{p} d^4 + c_2 d^3)(p_1 - p_2)/L \qquad (\text{ergs s}^{-1}) \tag{5.16}$$

where

$$c_1 = \pi/128 \tag{5.17}$$

$$c_2 = \pi/16[(\pi/2)(kT/m)]^{1/2}(2 - f)/f \tag{5.18}$$

and $f$ is the fraction of molecules that are adsorbed and reemitted at the wall

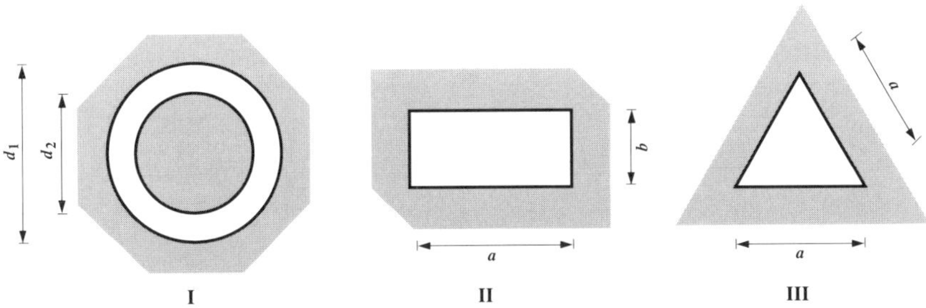


**Figure 5.6** Cross sections of I—a circular annulus (region between two concentric pipes of circular cross section), II—a rectangular channel, III—a triangular (equilateral) channel.

of the pipe and $1 - f$ represents the fraction of molecules that are specularly reflected.

When the average pressure $\bar{p}$ is such that $c_1 \bar{p} d^4 = c_2 d^3$, the flow rate departs from that predicted by Eqs. 5.11 and 5.12. The pressure at which the terms in $d^4$ and $d^3$ in Eq. 5.12 become equal is termed the *transition pressure* $p_{tr}$. The value of the transition pressure is

$$p_{tr} = c_2 / c_1 d \tag{5.19}$$

### 5.3.5 Laminar Conductance of a Circular Annulus

The conductance of the region between two concentric straight pipes of uniform circular cross section (Fig. 5.6, diagram I) (Dushman, 1966, p. 86) is

$$C_l = (\pi/8\eta)(\bar{p}/L)\left[r_1^4 - r_2^4 - \left(r_1^2 - r_2^2\right)^2/\ln(r_1/r_2)\right] \quad (\text{cm}^3\ \text{s}^{-1}) \tag{5.20}$$

with $\eta$ (poise), $\bar{p}$ (dyn cm$^{-2}$), $L$ (cm), $r_1$ (cm) radius of the external pipe, and $r_2$ (cm) radius of the core.

For air at $T = 23°\text{C}$

$$C_l = 2956(\bar{p}/L)\left[r_1^4 - r_2^4 - \left(r_1^2 - r_2^2\right)/\ln(r_1/r_2)\right] \quad (\text{liters s}^{-1}) \tag{5.20a}$$

with $\bar{p}$ (Torr), $L$ (cm), and $r_1$, $r_2$ (cm).

## 5.4 *Laminar Flow through Long Channels*

This type of flow is denoted by subscript l.

**Table 5.1**
Values of Correction Factor Y in Eq. 5.22 for Different Ratios $a$ (Width)/$b$ (Thickness) of a Channel of Rectangular Cross Section[a]

| $a/b$ | 1.00 | 0.9 | 0.8 | 0.7 | 0.6 | 0.5 | 0.4 | 0.3 | 0.2 | 0.1 |
|---|---|---|---|---|---|---|---|---|---|---|
| Y | 1.00 | 0.99 | 0.98 | 0.95 | 0.90 | 0.82 | 0.71 | 0.58 | 0.42 | 0.23 |

[a]After R. Loevinger, in Guthrie and Wakelring (1949, p. 32).

### 5.4.1 Laminar Conductance of Channels of Elliptical Cross Section

$$C_l = (\pi \bar{p}/4\eta L)[a^3b^3/(a^2 + b^2)] \qquad (\text{cm}^3\ \text{s}^{-1}) \qquad (5.21)$$

with $\bar{p}$ (dyn cm$^{-2}$), $\eta$ (poise), $L$ (cm) the length of the channel, $a$ (cm) the semimajor axis, and $b$ (cm) the semiminor axis (Dushman, 1966, p. 87).

For air at $T = 23°\text{C}$

$$C_l = 5912(\bar{p}/L)[a^3b^3/(a^2 + b^2)] \qquad (\text{liters s}^{-1}) \qquad (5.21\text{a})$$

with $\bar{p}$ (Torr), $L$ (cm), and $a$, $b$ (cm).

### 5.4.2 Laminar Conductance of Channels of Rectangular Cross Section

The conductance of channels [given by R. Loevinger in Guthrie and Wakelring (1949, p. 32) (see also Fig. 5.6, diagram II)] is, for air at $T = 23°\text{C}$

$$C_l = 267Ya^2b^2\bar{p}/L \qquad (\text{liters s}^{-1}) \qquad (5.22)$$

with $a$ (cm), $b$ (cm) the sides of the duct, $\bar{p}$ (Torr), and $L$ (cm) and where $Y = f(a/b)$ is the correction factor. The values of $Y$ are listed in Table 5.1.

### 5.4.3 Laminar Conductance of Channels of Triangular (Equilateral) Cross Section

Refer to Fig. 5.6, diagram III. For air at $T = 23°\text{C}$

$$C_l = 32.20\bar{p}a^4/L \qquad (\text{liters s}^{-1}) \qquad (5.23)$$

with $\bar{p}$ (Torr), $a$ (cm) the side of the triangle, $L$ (cm) the length of the channel.

## *5.5 Laminar Flow through Short Pipes of Uniform Circular Cross Section*

A pipe is considered *short* when its conductance to gas flow compares to the conductance of the aperture of the pipe. Short pipes lead to chocked flow.

For the equations of viscous flow in short pipes with either a chocked or nonchocked exit, see Santeler (1986). The viscous flow in short rectangular channels has been analyzed by Kieser and Grundner (1980), O'Hanlon (1987), and Sasaki and Yasunaga (1982).

## *Exercises*

**Exercise 5.1**

Calculate the nonchocked laminar flow rate of gas through an aperture for (a) monoatomic, (b) diatomic, and (c) polyatomic gases at $T = 23°C$.

**Solution 5.1**

Use Eq. 5.1:

$$Q_{lnc} = \left[(2\nu/\nu - 1)(RT_1/M)^{1/2}(p_2/p_1)^{1/\nu}\right]^{1/2} \times \left[1 - (p_2/p_1)^{(\nu-1)/\nu}\right]^{1/2} p_1 A$$

Substitute

$R = 8.314 \times 10^7$ ergs mol$^{-1}$ K$^{-1}$ (see Table 1.3)
$T_1 = 273 + 23 = 296$ K
$M_a$ (g mol$^{-1}$) for monoatomic gases
$M_b$ (g mol$^{-1}$) for diatomic gases
$M_c$ (g mol$^{-1}$) for polyatomic gases
$\nu_a$ 1.67 dimensionless (see Table 3.14)
$\nu_b$ 1.40 dimensionless (see Table 3.14)
$\nu_c$ 1.33 dimensionless (see Table 3.14)
$p_1, p_2$ (dyn cm$^{-2}$)
A (cm)$^2$

Thus

(a) For monoatomic gases:

$$Q_{lnc} = \left[(2 \times 1.67/1.67 - 1)\left(8.314 \times 10^7 \text{ ergs mol}^{-1} \text{ K}^{-1} \times 296 \text{ K}/M_a \text{ g mol}^{-1}\right)\right]^{1/2} \times \left\{(p_2/p_1)^{1/1.67}[1 - (p_2/p_1)]^{(1.67-1)/1.67)}\right\}^{1/2} p_1 \text{ dyn cm}^{-2} \text{ A cm}^2$$

$$= \left(3.5 \times 10^5/M_a^{1/2}\right)(p_2/p_1)^{0.599}\left[1 - (p_2/p_1)^{0.401}\right]^{1/2} p_1 A \quad (\text{ergs s}^{-1})$$

(b) For diatomic gases:

$$Q_{\text{lnc}} = \Big[(2 \times 1.40/1.40 - 1)\big(8.314 \times 10^7 \text{ ergs mol}^{-1}\text{ K}^{-1} \times 296 \text{ K}/M_b \text{ g mol}^{-1}\big)\Big]^{1/2}$$
$$\times \Big\{(p_2/p_1)^{1/1.40}\big[1 - (p_2/p_1)\big]^{(1.40-1)/1.40}\Big\}^{1/2} p_1 \text{ dyn cm}^{-2} \times A \text{ cm}^2$$
$$= \big(4.15 \times 10^5/M_b^{1/2}\big)(p_2/p_1)^{0.714}\Big[1 - (p_2/p_1)^{0.286}\Big]^{1/2} p_1 A \quad (\text{ergs s}^{-1})$$

(c) For polyatomic gases:

$$Q_{\text{lnc}} = \Big[(2 \times 1.33/1.33 - 1)\big(8.314 \times 10^7 \text{ ergs mol}^{-1}\text{ K}^{-1} \times 296 \text{ K}/M_c \text{ g mol}^{-1}\big)\Big]^{1/2}$$
$$\times \Big\{(p_2/p_1)^{1/1.33}\big[1 - (p_2/p_1)\big]^{(1.33-1)/1.33}\Big\}^{1/2} p_1 \text{ dyn cm}^{-2} \times A \text{ cm}^2$$
$$= \big(4.45 \times 10^5/M_c^{1/2}\big)(p_2/p_1)^{0.752}\Big[1 - (p_2/p_1)^{0.248}\Big]^{1/2} p_1 A \quad (\text{ergs s}^{-1})$$

**Exercise 5.2**

Calculate the ratio of nonchocked laminar flow rates of air and hydrogen at $T = 23°\text{C}$ through a small aperture. The inlet and outlet pressures from the aperture are the same for both gases.

**Solution 5.2**

Both air and hydrogen are diatomic gases; thus $\nu_{\text{air}} = \nu_{H_2} = 1.40$. Write Eq. 5.1 for both gases and divide the expressions obtained. Thus

$$Q_{\text{lnc air}}/Q_{\text{lnc } H_2} = \big(M_{H_2}/M_{\text{air}}\big)^{1/2}$$

Substitute

$M_{\text{air}} = 28.98 \times 10^{-3}$ kg mol$^{-1}$ (see Table 1.2)
$M_{H_2} = 2.0158 \times 10^{-3}$ kg mol$^{-1}$ (see Table 1.2)

Then

$$Q_{\text{lnc air}}/Q_{\text{lnc } H_2} = \big(2.0158 \times 10^{-3} \text{ kg mol}^{-1}/28.98 \times 10^{-3} \text{ kg mol}^{-1}\big)^{1/2}$$
$$= 0.26$$

**Exercise 5.3**

Calculate the laminar flow conductance of a straight pipe, 3 m long of 1.6-cm inner diameter, for air at $T = 0°C$ and $T = 23°C$, assuming that the pressures at the inlet and outlet from the pipe are 100 mbar and 80 mbar, respectively.

**Solution 5.3**

Use Eq. 5.12:

$$C_l = (\pi d^4/128\eta L)\bar{p}$$

Substitute

$$
\begin{aligned}
d &= 1.6 \text{ cm} \\
\eta_{\text{air }23°\text{C}} &= 177.7 \times 10^{-6} \text{ poise (see Table 3.12)} \\
\eta_{\text{air }0°\text{C}} &= 170.7 \times 10^{-6} \text{ poise (see Table 3.12)} \\
L &= 300 \text{ cm} \\
\bar{p} &= (p_1 + p_2)/2 = (100 \text{ mbar} + 80 \text{ mbar})/2 = 90 \text{ mbar} \\
&= 90 \text{ mbar} \times 1000 \text{ dyn cm}^{-2}/\text{mbar} \\
&= 8.0 \times 10^4 \text{ dyn cm}^{-2}
\end{aligned}
$$

For air at $T = 23°C$:

$$C_l = (3.14 \times 1.6^4/128 \times 177.7 \times 10^{-6} \times 300) \times 9.0 \times 10^4$$

$$= 271415 \text{ cm}^3 \text{ s}^{-1} = 271.415 \text{ liters s}^{-1}$$

For air at $T = 0°C$:

$$C_l = (3.14 \times 1.6^4/128 \times 170.7 \times 10^{-6} \times 300) \times 9.0 \times 10^4$$

$$= 282545 \text{ cm}^3 \text{ s}^{-1} = 282.545 \text{ liters s}^{-1}$$

**Exercise 5.4**

Express the volume rate of gas flow at a specified temperature in terms of the pressure gradient in a straight pipe of uniform circular cross section.

**Solution 5.4**

Write Eqs. 4.6b and 5.11a

$$Q = \bar{p}V' \quad (\text{at a specified } T)$$

$$Q = (\pi d^4/128\eta L)\bar{p}(p_1 - p_2)$$

and solve the system for $V'$

$$V' = (\pi d^4/128\eta L)(p_1 - p_2) \quad (\text{cm}^3 \text{ s}^{-1} \text{ at a specified } T)$$

with $d$ (cm), $L$ (cm), $\eta$ (poise), $p_1$, $p_2$ (dyn cm$^{-2}$).

For $T = 23°C$, $\eta = 177.7 \times 10^{-6}$ poise (see Table 3.12). Thus

$$V' = (\pi/128 \times 177.7 \times 10^{-6})(d^4/L)(p_1 - p_2)$$
$$= 138(d^4/L)(p_1 - p_2) \qquad (\text{liters s}^{-1})$$

with $d$ (cm), $L$ (cm), $p_1$, $p_2$ (mbar).

Note that $V'$ is the volume rate of flow at a specified temperature *in the middle of the pipe*, where the pressure is $\bar{p}$. Since the pressure drop along the pipe is usually small compared to $\bar{p}$, the difference between the values of $V'$ in the middle of the pipe and at either end can be disregarded.

**Exercise 5.5**

Find the relationship permitting evaluation of whether a pipe of length $L$ and inner diameter $d$, connected to a pump of a certain pumping speed, permits the efficient utilization of the pump when air at $T = 23°C$ is pumped.

**Solution 5.5**

It is experimentally proved that the pressure drop along a pipe connecting a pump to a vessel must be less than (or at least one-fifth of) the pressure at the port of the pump at operating level. With that in mind, use Eqs. 4.38a, 4.31b, and 5.12b written as

(i) $Q = p_P S_P$
(ii) $Q = (p_1 - p_P)C$
(iii) $C_l = 138(d^4/L)\bar{p}$ with $\bar{p}$ (mbar) (liters $s^{-1}$)

where
$p_P$ = the pressure at the port of the pump
$p_1$ = the pressure at the utilization point
$S_P$ = the pumping speed at the port of the pump at operating level
$Q$ = the throughput of the pipe
$C$ = the conductance of the pipe
$C_l$ = the conductance of a straight pipe of uniform cross section for the laminar flow of gas
$d$ = the diameter of the pipe
$L$ = the length of the pipe
$\eta$ = the viscosity of gas flowing through the pipe at a specified temperature
$\bar{p}$ = the average pressure

Write (ii) as

$$p_1 - p_P = Q/C$$

and substitute $C$ by its expression (iii):

$$p_1 - p_P = QL/138d^4\bar{p}$$

Substitute $Q$ by its expression (ii) and rearrange the variables as

(iv) $$p_1 - p_P = (S_P L/138 d^4)(p_P/\bar{p})$$

Since $\bar{p} = (p_1 + p_P)/2$ the term $p_P/\bar{p}$ in (iv) becomes

$$p_P/\bar{p} = 2p_P/(p_1 + p_P) = 2/[(p_1/p_P) + 1]$$

According to the condition $p_1 - p_P \leqslant 1.5 p_P$

$$p_1/p_P = 1.2$$

Thus

$$2/[(p_1/p_P) + 1] = 2/1.2 + 1 = 0.9$$

It can be seen that if $p_1 - p_P$ is much less than $(1/5)p_P$, the expression $(p_P/\bar{p}) \simeq 1$ and (iv) becomes

$$p_1 - p_P = 7.2 \times 10^{-3} S_P L/d^4$$

which is Eq. 5.15a.

**Exercise 5.6**

A double-stage rotary pump with a pumping speed of 30 $m^3\ h^{-1}$ (theoretical displacement) without gas ballast is connected by a pipe 80 cm long to a vessel where pressure must be maintained at $10^{-2}$ mbar. Evaluate the diameter of the pipe.

**Solution 5.6**

Use Eq. 5.6:

$$p_1 - p_P = 7.2 \times 10^{-3} S_P L/d^4$$

Substitute

$p_P = 10^{-2}$ mbar
$p_1 - p_P = (1/5)p_P = 10^{-2}$ mbar$/5 = 2 \times 10^{-3}$ mbar (see Section 5.3.3)
$S_P \simeq 20$ $m^3\ h^{-1}$ (value found on the performance curve of the pump)
$L = 80$ cm

Thus

$$d = \left[(7.2 \times 10^{-3} \times 20 \times 80)/2 \times 10^{-3}\right]^{1/4} = 8.71 \text{ cm}$$

**Exercise 5.7**

Calculate the laminar flow conductance of a long pipe, $L > 20d$, of uniform circular cross section for $H_2$, $H_2O$ (vapor), $CO_2$, CO, $N_2$, He, and Ne at $T = 23°C$. For calculations use the laminar flow conductance for air at $T = 23°C$. Comment on the results.

**Solution 5.7**

Use Eq. 5.12

$$C_l = (\pi d^4/128\eta L)\bar{p}$$

written for air (i) and one of the gases, e.g., $H_2$ (ii)

(i) $C_{l\,\text{air}} = (\pi d^4/128\eta_{\text{air at 23°C}}L)\bar{p}$
(ii) $C_{l\,H_2} = (\pi d^4/128\eta_{H_2\text{ at 23°C}}L)\bar{p}$

Divide (i) by (ii):

$$C_{l\,H_2\text{ at 23°C}} = (\eta_{H_2\text{ at 23°C}}/\eta_{\text{air at 23°C}})C_{l\,\text{air at 23°C}}$$

The results of calculations are listed in the following table.

| Gas | $\eta_{23°C}$[a] ($\mu$ poise) | Air/Gas $\eta_{23°C}/\eta_{23°C}$ | $C_{l\,\text{gas at 23°C}}$ (e.g., liters $s^{-1}$) |
|---|---|---|---|
| Air | 177.7 | 1.00 | $1.00C_{l\,\text{air at 23°C}}$ |
| $H_2$ | 92.3 | 1.92 | $1.92C_{l\,\text{air at 23°C}}$ |
| $H_2O$ vapor | 96.9 | 1.83 | $1.83C_{l\,\text{air at 23°C}}$ |
| $CO_2$ | 143.3 | 1.24 | $1.24C_{l\,\text{air at 23°C}}$ |
| CO | 172.2 | 1.03 | $1.03C_{l\,\text{air at 23°C}}$ |
| $N_2$ | 173.2 | 1.03 | $1.03C_{l\,\text{air at 23°C}}$ |
| He | 192.4 | 0.92 | $0.92C_{l\,\text{air at 23°C}}$ |
| Ne | 330.7 | 0.54 | $0.54C_{l\,\text{air at 23°C}}$ |

[a]See Table 3.12.

The laminar flow conductance of a long pipe of uniform circular cross section is practically the same for air, CO, and $N_2$ all at the same temperature.

**Exercise 5.8**

A vacuum vessel (1, Fig. 5A) that contains air at $T = 23°C$ is roughed by a rotary-vane pump (3) until reaching the pressure at which the diffusion pump (2) can take over the evacuation process.

The diffusion pump (2), which has a pumping speed of 940 liters $s^{-1}$ at the operation level of $10^{-4}$ mbar, requires a backing pressure (limiting forepressure) of at least $2 \times 10^{-1}$ mbar. Because of space limitations the rotary-vane pump (3) must be located 0.95 m away form the diffusion pump.

Calculate the diameter of the backing line and the pumping speed of the rotary-vane pump, assuming steady flow conditions (i.e., $Q$ = const., $T$ = const., absence of gas load).

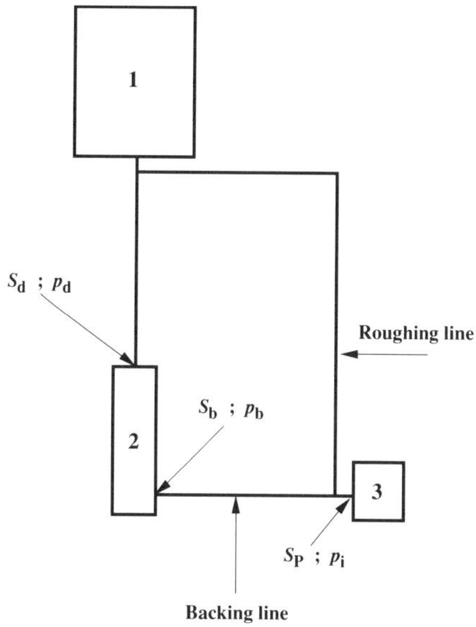


Figure 5A

## Solution 5.8

The basic considerations in the design of a vacuum system (see Section 4.13) provide an adequate basis for the following calculations; i.e., the flow rate of gas through the system is conservative. Thus, the backing pump must handle the same throughput as the diffusion pump.

To determine the throughput of the diffusion pump, use Eq. 4.38a written as

$$Q_d = S_d p_d$$

where $Q_d$ = throughput of the diffusion pump
$S_d$ = 940 liters $s^{-1}$, pumping speed of the diffusion pump
$p_d$ = $10^{-4}$ mbar, pressure at the port of the diffusion pump

Hence

$$Q_d = 940 \text{ liters s}^{-1} \times 10^{-4} \text{ mbar} = 9.4 \times 10^{-2} \text{ mbar liters s}^{-1}$$

$$(\text{at } T = 23°\text{C})$$

The rotary-vane pump must handle the same throughput. Using the same equation (Eq. 4.38a) for the backing pump it follows that

$$S_b = Q_d/p_b$$

where $S_b$ = pumping speed required at the backing port of the diffusion pump
$Q_d$ = $9.4 \times 10^{-2}$ mbar liters $s^{-1}$ (at $T = 23°C$) air-flow rate through diffusion pump
$p_b$ = $2 \times 10^{-1}$ mbar backing pressure

Thus

$$S_b = 9.4 \times 10^{-2} \text{ mbar liters s}^{-1}/2 \times 10^{-1} \text{ mbar} = 0.5 \text{ liters s}^{-1}$$

To keep losses of the pumping speed through the backing line below 20%, take the pumping speed $S_P$ at the port of the backing pump (3) (first approximation):

$$S_P = 10S_b = 10 \times 0.5 \text{ liters s}^{-1} = 5 \text{ liters s}^{-1}$$

Again, from Eq. 4.38 written as

$$p_i = Q_d/S_P$$

where $p_i$ = the pressure at the port of the backing pump (3)
$Q_d$ = $9.4 \times 10^{-2}$ mbar liters $s^{-1}$ (at $T = 23°C$)
$S_P$ = 5 liters $s^{-1}$ (see above)

there results

$$p_i = 9.4 \times 10^{-2} \text{ mbar liters s}^{-1}/5 \text{ liters s}^{-1} = 1.88 \times 10^{-2} \text{ mbar}$$

Now evaluate the diameter of the pipe connecting the diffusion pump to the rotary-vane pump (backing line). Use Eq. 5.15a written as

$$p_b - p_i = 7.2 \times 10^{-3} S_p L/d^4$$

For $p_b - p_i = (1/5)p_i$. It follows that $p_b - p_i = 1/5 \times 1.88 \times 10^{-2}$ mbar $= 3.76 \times 10^{-3}$ mbar.

Substitute

$p_b - p_i = 3.76 \times 10^{-3}$ mbar
$S_P = 5$ liters $s^{-1}$
$L = 95$ cm

Therefore

$$d^4 = (7.2 \times 10^{-3}/3.76 \times 10^{-3}) \times 5 \times 95 = 909.6 \text{ cm}^4$$

$$d = 5.5 \text{ cm}$$

Choose for $d$ the next standardized size, say, $d = 6$ cm.

A check of the flow regime of air through the backing line shows that, according to Eq. 4.20, $Q_l < 143d$, and with $Q_l = Q_d = 9.4 \times 10^{-2}$ mbar liters s$^{-1}$ (at $T = 23°C$), $9.4 \times 10^{-2}$ mbar liters s$^{-1}$ $< 143 \times 6$. Obviously, the flow is laminar.

To avoid the stopdown of the diffusion pump resulting from the crossover pressure, the backing pump must have a pumping speed much larger than $S_P = 5$ liters s$^{-1}$. Consequently, one may select, in the range of standardized rotary-vane pumps, the next size, viz., $S_P = 17$ liters s$^{-1}$.

The resulting net pumping speed at the backing port of the diffusion pump can be calculated, for the new conditions, by using Eq. 4.41 written as

$$S_b = S_P C_l/(S_P + C_l)$$

where

$$S_P = 17 \text{ liters s}^{-1}$$

$$C_l = 138(d^4/L)\bar{p} \qquad \text{(Eq. 5.12b)}$$

Substitute

$d = 6$ cm
$L = 95$ cm
$\bar{p} = (p_b - p_i)/2 = (2 \times 10^{-1} \text{ mbar} + 1.88 \times 10^{-2} \text{ mbar})/2 = 0.109$ mbar

Thus

$$C_l = 138 \times (6^4/95) \times 0.109 = 205 \text{ liters s}^{-1}$$

$$S_b = 17 \text{ liters s}^{-1} \times 105 \text{ liters s}^{-1}/(17 \text{ liters s}^{-1} + 205 \text{ liters s}^{-1})$$

$$= 15.7 \text{ liters s}^{-1}$$

The backpressure will be, according to Eq. 4.38a

$$p_b = Q_d/S_b = 9.4 \times 10^{-2} \text{ mbar liters s}^{-1}/17.5 \text{ liters s}^{-1}$$

$$= 5.37 \times 10^{-3} \text{ mbar}$$

which is less than the backing pressure $2 \times 10^{-1}$ mbar.

**Exercise 5.9**

Compare the laminar flow conductances for air at $T = 23°C$ of long, straight (a) pipes of uniform circular cross section, (b) channels of square cross section, and (c) channels of triangular (equilateral) cross section.

**Solution 5.9**

To compare the laminar flow conductances requested, assume that the pipe of uniform circular cross section and the channels of square and triangular

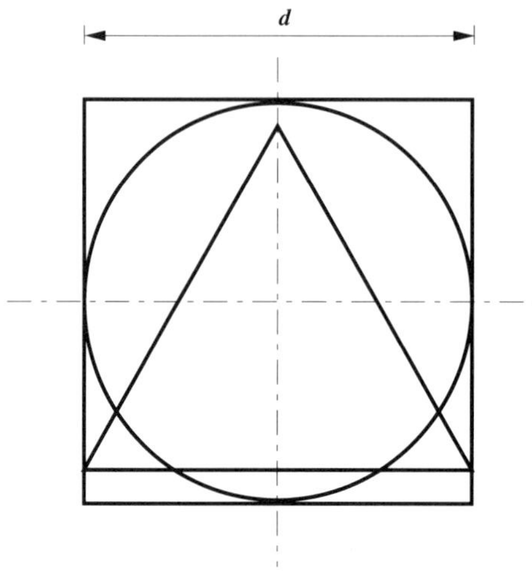

**Figure 5B**

(equilateral) cross section have the same length $L$ and the sides of the channels are equal to the diameter of the pipe. (See Fig. 5B.)

Use Eqs. 5.12c, 5.22, and 5.23 and subscripts ○, □, and △ to designate the cross sections.

(a) Circular cross section:

$$C_{1\bigcirc} = 184(d^4/L)\bar{p} \qquad \text{air at } T = 23°\text{C} \qquad (\text{liters s}^{-1})$$

with $d$ (cm), $L$ (cm), $\bar{p}$ (Torr).

(b) Square cross section:

$$C_{1\square} = 267Ya^2b^2\bar{p}/L \qquad \text{air at } T = 23°\text{C} \qquad (\text{liters s}^{-1})$$

$$= 267d^4\bar{p}/L$$

with $Y = 1$ since $a = b$ (see Table 5.1), $a = b = d$, $d$ (cm), $L$ (cm), $\bar{p}$ (Torr).

(c) Triangular (equilateral) cross section:

$$C_{1\triangle} = 32.20a^4\bar{p}/L \qquad \text{air at } T = 23°\text{C} \qquad (\text{liters s}^{-1})$$

$$= 32.00d^4\bar{p}/L$$

with $d$ (cm), $L$ (cm), $\bar{p}$ (Torr).

Thus

$$C_{1\square}/C_{1\bigcirc} = \left(267d^4\bar{p}/L\right)/184\left(d^4/L\right)\bar{p}$$

$$= 1.45 \qquad \text{for air at } T = 23°\text{C}$$

$$C_{1\triangle}/C_{1\bigcirc} = \left(32.20d^4\bar{p}/L\right)/184\left(d^4/L\right)\bar{p}$$

$$= 0.175 \qquad \text{for air at } T = 23°\text{C}$$

**Exercise 5.10**

Calculate the ratio of chocked laminar flow conductance through an aperture of area A ($cm^2$) of diatomic gases of different molar masses, at the same temperature $T_1$. Assume application for $H_2$ and air crossing an aperture of diameter $5.0 \times 10^{-3}$ cm.

**Solution 5.10**

Use Eq. 5.8

$$C_{lch} = O_{ch}\left\{(\nu RT_1/M)\Big/\left[2/(\nu + 1)^{(\nu+1)/(\nu-1)}\right]\right\}^{1/2} A$$

and subscripts 1 and 2 for gases of different molar masses. Divide the resulting equations.

Thus

$$C_{lch1}/C_{lch2} = (1/M_1)^{1/2}\left\{(\nu RT_1)\Big/\left[2/(\nu + 1)^{(\nu+1)/(\nu-1)}\right]\right\}^{1/2} A/$$

$$(1/M_2)^{1/2}\left\{(\nu RT_1)\Big/\left[2/(\nu + 1)^{(\nu+1)/(\nu-1)}\right]\right\}^{1/2} A$$

$$= (M_2/M_1)^{1/2}$$

Substitute

$M_1 = M_{H_2} = 2.0 \times 10^{-3}$ kg mol$^{-1}$ (see Table 1.2)
$M_2 = M_{air} = 28.98 \times 10^{-3}$ kg mol$^{-1}$ (see Table 1.2)

Thus

$$C_{lch\,H_2}/C_{lch\,air} = 2.0 \times 10^{-3}\ \text{kg mol}^{-1}/28.98 \times 10^{-3}\ \text{kg mol}^{-1}$$

$$= 6.9 \times 10^{-2}$$

## References

Dushman, S. (1966), in *Scientific Foundations of Vacuum Technique*, 2nd ed., J. M. Lafferty (ed.), Wiley, New York.

Guthrie, A., and Wakelring, R. K. (1949), *Vacuum Equipment and Techniques*, McGraw-Hill, New York.

Kieser, J., and Grundner, M. (1980), *Le Vide, 201*, 376.

O'Hanlon, J. F. (1987), *J. Vac. Sci. Technol.* A5, *1*, 98.

Santeler, J. D. (1986), *J. Vac. Sci. Technol.* A4, *3*, 348.

Santeler, J. D., Holkeboer, H. D., Jones, W. D., and Pagano, F. (1966), *Vacuum Technology and Space Simulation*, National Aeronautics and Space Administration, NASA SP-105.

Sasaki, S., and Yasunaga, S. (1982), *J. Vac. Sci. Technol.* *25*, 157.

Van Atta, C. M. (1965), *Vacuum Science and Engineering*, McGraw-Hill, New York.

# 6

# Steady Flow of Gas in the Molecular Range

## 6.1 *Molecular Flow through an Aperture*

Molecules from two large chambers at very low pressures connected by an aperture will pass from one chamber to the other, in both directions, without making any collisions in the neighborhood of the aperture, if conditions are conducive. In contrast with the laminar flow of gas, no mass motion of the molecules will develop and the gas will be practically unaffected by the presence of the aperture.

Molecular flow through an aperture (denoted by subscript ma) occurs when the mean free path of the gas species ($\lambda$) is very large compared to the radius ($r$) of the aperture, i.e., $\lambda \gg r$, and the wall of the flat plate in which the aperture is cut is vanishingly small in the region of the aperture, so that it has no length ideally. Such construction is termed *thin-edged aperture* (Fig. 6.1).

### 6.1.1 Molecular Throughput of an Aperture

$$Q_{ma} = (1/4)(8kT/\pi m)^{1/2}(p_1 - p_2)A \qquad (\text{erg s}^{-1}) \qquad (6.1)$$

$$= (1/4)(8RT/\pi M)^{1/2}(p_1 - p_2)A \qquad (6.1a)$$

$$= 3.64 \times 10^3 (T/M)^{1/2}(p_1 - p_2) \qquad (\text{dyn cm}^{-2}\ \text{cm}^{-3}\ \text{s}^{-1}) \quad (6.1b)$$

with $T$ (K), M (g mol$^{-1}$), $p_1$, $p_2$ (dyn cm$^{-2}$), and A (cm$^2$) area of the aperture.

Equations 6.1–6.1b hold true for $p_1 > 10p_2$.

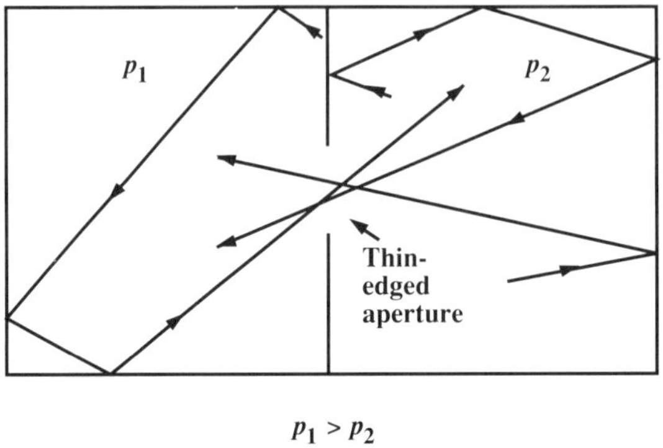


**Figure 6.1** Schematic of molecular flow of gas through a "thin"-edged aperture.

### 6.1.2 Molecular Conductance of an Aperture

$$C_{ma} = Q/(p_1 - p_2) \quad \text{(see Eq. 4.31b)} \quad (\text{liters s}^{-1}) \tag{6.2}$$

$$= (1/4)(8kT/\pi m)^{1/2} A \quad (\text{cm}^3\ \text{s}^{-1}) \tag{6.2a}$$

$$= (1/4)(8RT/\pi M)^{1/2} A \quad (\text{cm}^3\ \text{s}^{-1}) \tag{6.2b}$$

$$= 3.64(T/M)^{1/2} A \quad (\text{liters s}^{-1}) \tag{6.2c}$$

with $T$ (K), M (g mol$^{-1}$), and A (cm$^2$) area of the aperture.

For air at $T = 23°\text{C}$

$$C_{ma} = 11.6A \quad (\text{liters s}^{-1}) \tag{6.2d}$$

*The molecular flow conductance of an aperture depends only on the species* (M) *and the temperature* ($T$) *of gas and is independent of the gas pressure.*

Table 6.1 lists the values of the molecular conductance $C_{ma}$ of a circular aperture of diameter $d$ for air and various gases at $T = 23°\text{C}$. Note that for constant temperature, $C_{ma}$ varies with the molecular mass M of gas.

### 6.1.3 Molecular Pumping Speed of an Aperture

$$S_{ma} = C_{ma}((p_1 - p_2)/p_1) \quad (\text{e.g., liters s}^{-1}) \tag{6.3}$$

$$= 3.64(T/M)^{1/2} A(1 - p_2/p_1) \quad (\text{liters s}^{-1}) \tag{6.3a}$$

with $T$ (K), M (g mol$^{-1}$), and $p_1$, $p_2$ (mbar or Torr).

**Table 6.1**

Molecular Conductance $C_{ma}$ (liters $s^{-1}$) of Circular Apertures of Diameter $d$ for Air and Various Gases [with molar masses $M$ (g $mol^{-1}$)[a] as given] at $T = 23°C$, Calculated from Eq. 6.2c

| $d$ (cm) | $H_2$ 2.0158 | He 4.0026 | Ne 20.183 | $N_2$ 28.0134 | Air 28.98 | $O_2$ 31.9988 | Ar 35.948 | Kr 83.80 | Xe 131.30 |
|---|---|---|---|---|---|---|---|---|---|
| 0.1 | $3.46 \times 10^{-1}$ | $2.46 \times 10^{-1}$ | $1.09 \times 10^{-1}$ | $9.29 \times 10^{-2}$ | $9.13 \times 10^{-2}$ | $8.70 \times 10^{-2}$ | $7.78 \times 10^{-2}$ | $5.37 \times 10^{-2}$ | $4.29 \times 10^{-2}$ |
| 0.2 | 1.39 | $9.83 \times 10^{-1}$ | $4.38 \times 10^{-1}$ | $3.72 \times 10^{-1}$ | $3.66 \times 10^{-1}$ | $3.48 \times 10^{-1}$ | $3.11 \times 10^{-1}$ | $2.15 \times 10^{-1}$ | $1.72 \times 10^{-1}$ |
| 0.3 | 3.12 | 2.21 | $9.85 \times 10^{-1}$ | $8.36 \times 10^{-1}$ | $8.22 \times 10^{-1}$ | $7.82 \times 10^{-1}$ | $7.00 \times 10^{-1}$ | $4.83 \times 10^{-1}$ | $3.86 \times 10^{-1}$ |
| 0.4 | 5.54 | 3.93 | 1.75 | 1.49 | 1.46 | 1.39 | 1.24 | $8.6 \times 10^{-1}$ | $6.87 \times 10^{-1}$ |
| 0.5 | 8.66 | 6.14 | 2.74 | 2.32 | 2.28 | 2.17 | 1.94 | 1.34 | 1.07 |
| 0.6 | 12.47 | 8.85 | 3.94 | 3.34 | 3.29 | 3.13 | 2.80 | 1.93 | 1.54 |
| 0.7 | 16.97 | 12.04 | 5.36 | 4.35 | 4.47 | 4.26 | 3.81 | 2.63 | 2.10 |
| 0.8 | 22.16 | 15.73 | 7.00 | 5.94 | 5.84 | 5.56 | 4.98 | 3.44 | 2.75 |
| 0.9 | 28.04 | 19.90 | 8.86 | 7.52 | 7.40 | 7.04 | 6.30 | 4.35 | 3.48 |
| 1.0 | 34.6 | 24.6 | 10.9 | 9.29 | 9.13 | 8.70 | 7.78 | 5.37 | 4.29 |

[a]See Table 1.2.

For air at $T = 23°C$

$$S_{ma} = 11.6A(1 - p_2/p_1) \qquad (\text{liters s}^{-1}) \tag{6.3b}$$

when $p_2 \leqslant 10p_1$

$$S_{ma} = 11.6A \qquad (\text{liters s}^{-1}) \tag{6.3c}$$

For the molecular flow through a circular aperture located coaxially in a disk across a cylindrical pipe, see Bureau et al. (1952).

## 6.2 Molecular Flow through a Diaphragm—Diaphragm Effect

Refer to Fig. 6.2. The diaphragm effect (denoted by subscript md) occurs when a very large vessel (1) is connected through a pipe (2) of cross-sectional area $A_0$ to another large vessel (3), through an aperture of cross-sectional area $A$, and the pressure in the vessels (1, 3) is very low.

The aperture cut in the wall of volume (3) is to be considered small when approached from the right, but of the same order of magnitude as $A_0$ when approached from the left.

### 6.2.1 Molecular Conductance of a Diaphragm

$$C_{md} = C_{ma}/[1 - (A/A_0)] \qquad (\text{e.g., liters s}^{-1}) \tag{6.4}$$

with $C_{ma}$ (e.g., liters $s^{-1}$; see Eq. 6.3), $A_0$ (e.g., $cm^2$), and $A$ (e.g., $cm^2$). For air at $T = 23°C$:

$$C_{md} = 11.6A/[1 - (A/A_0)] \qquad (\text{liters s}^{-1}) \tag{6.4a}$$

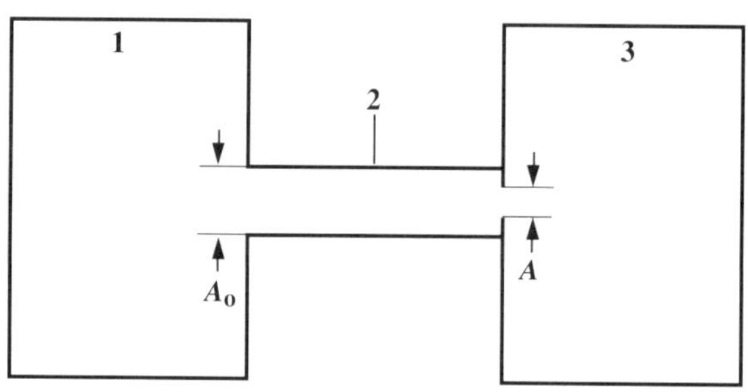


**Figure 6.2** Schematic of the configuration leading to the diaphragm effect: 1, 3—very large volumes; 2—pipe of cross-sectional area $A_0$; A—cross-sectional area of the aperture cut in the wall of the volume (3).

For $A/A_0 = 0.5$:

$$C_{md} = 2C_{ma} = 2 \times 11.6A \qquad (\text{liters s}^{-1}) \qquad (6.4b)$$

### 6.2.2 Molecular Pumping Speed of a Diaphragm

$$S_{md} = C_{md}[1 - (p_2/p_1)] \qquad (\text{e.g., liters s}^{-1}) \qquad (6.5)$$

For air at $T = 23°C$:

$$S_{md} = 11.6A[1 - (p_2/p_1)]/[1 - (A/A_0)] \qquad (\text{liters s}^{-1}) \quad (6.5a)$$

## 6.3 *Molecular Flow through Long Pipes*

At very low pressures the molecules move in a pipe (this type of flow is denoted by subscript mp) in random straight lines striking the wall of the pipe at the end of each flight. Calculations take into account only those molecules well inside the pipe far removed from the ends. Thus, the effect of the entrance (exit) aperture to the pipe is unimportant. Practically, a pipe is considered long when the length of the pipe $L \geqslant 20d$ (inner diameter of the pipe).

### 6.3.1 Molecular Throughput of a Pipe of Varying Cross Section and Perimeter (Knudsen, 1909, 1911)

$$Q_{mp} = (16/3)(RT/2\pi M)^{1/2}(p_1 - p_2)\Big/\int_0^L (o/A^2)\, dL \qquad (\text{ergs s}^{-1}) \qquad (6.6)$$

with $R$ (ergs mol$^{-1}$ K$^{-1}$), $T$ (K), M (g mol$^{-1}$), $p_1$, $p_2$ (dyn cm$^{-2}$), $o$ (cm) the perimeter of the aperture to the pipe, A (cm$^2$) cross-sectional area of the pipe, and $L$ (cm) the length of the pipe.

### 6.3.2 Molecular Conductance of a Pipe of Varying Cross Section and Perimeter (Knudsen, 1909, 1911)

$$C_{mp} = (16/3)(RT/2\pi M)^{1/2}\Big/\int_0^L (o/A^2)\, dL \qquad (\text{cm}^3\ \text{s}^{-1}) \quad (6.7)$$

where the symbols are as in Eq. 6.6.

*The molecular flow conductance of pipes is independent of pressure. For a certain species of gas* (M) *and temperature* ($T$), *the molecular flow conductance depends only on the geometry of the pipe.*

### 6.3.3 Molecular Conductance of a Pipe of Uniform Cross Section

$$C_{mp} = (16/3)(RT/2\pi M)^{1/2} A^2/oL \qquad (\text{cm}^3\ \text{s}^{-1}) \tag{6.8}$$

$$= 19.4(T/M)^{1/2} A^2/oL \qquad (\text{liters s}^{-1}) \tag{6.8a}$$

with $R$ (ergs mol$^{-1}$ K$^{-1}$), $T$ (K), M (g mol$^{-1}$), A (cm), $o$ (cm), and $L$ (cm). For a cylindrical pipe of inner diameter $d$ and length $L$:

$$C_{mp} = 3.81(T/M)^{1/2} d^3/L \qquad (\text{liters s}^{-1}) \tag{6.9}$$

For air at $T = 23°\text{C}$:

$$C_{mp} = 12.1d^3/L \qquad (\text{liters s}^{-1}) \tag{6.9a}$$

The molecular conductance for air at $T = 23°\text{C}$ of cylindrical pipes fitting standard con-flat flanges is illustrated in Fig. 6.3.

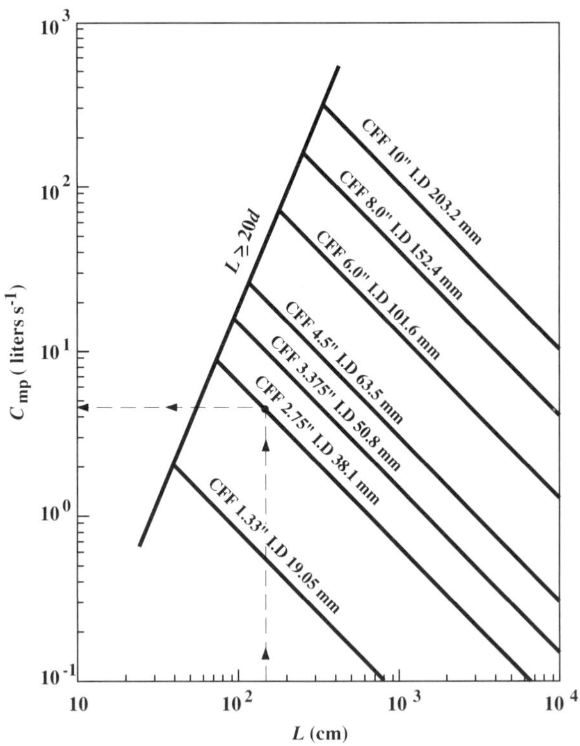


**Figure 6.3** Molecular flow conductance $C_{mp}$ for air at $T = 23°\text{C}$ of long cylindrical pipes fitting standard con-flat flanges calculated from Eq. 6.9a.

### 6.3.4 Molecular Conductance of a Tapered (Conical) Pipe of Circular Cross Section (Roth, 1990, p. 90)

$$C_{mp} = 7.62(T/M)^{1/2} d_1^2 d_2^2 / [(d_1 + d_2)L] \qquad (\text{liters s}^{-1}) \qquad (6.10)$$

with $T$ (K), M (g mol$^{-1}$), $d_1$, $d_2$ (cm) the diameters of the tapered pipe ends, and $L$ (cm) the length of the pipe.

For air at $T = 23°C$

$$C_{mp} = 24.35(d_1^2 d_2^2) / [(d_1 + d_2)L] \qquad (\text{liters s}^{-1}) \qquad (6.10a)$$

### 6.3.5 Molecular Conductance of a Circular Annulus

The conductance of the region between two concentric straight pipes of uniform circular cross section (Guthrie and Wakelring, 1949, p. 37) is

$$C_{mp} = (\pi/3)(RT/2\pi M)^{1/2}(d_1^2 - d_2^2)^2 K_0/(d_1 + d_2)L \qquad (\text{cm}^3\ \text{s}^{-1}) \qquad (6.11)$$

$$= 3.8(T/M)^{1/2}(d_1^2 - d_2^2)^2 K_0/(d_1 + d_2)L \qquad (\text{liters s}^{-1}) \qquad (6.11a)$$

with $R$ (ergs mol$^{-1}$ K$^{-1}$), M (g mol$^{-1}$), and $d^1$, $d^2$ (cm) (see Fig. 5.6, diagram I), $k_0 = f(d_2/d_1)$ correction factor for the asymmetry of the cross section as compared to that of a circle, and $L$ (cm) length of the concentric pipes. The values of the correction factor $K_0$ are listed in Table 6.2. For air at $T = 23°C$:

$$C_{mp} = 12.1(d_1^2 - d_2^2)^2 K_0/(d_1 + d_2)L \qquad (\text{liters s}^{-1}) \qquad (6.11b)$$

For further details concerning the calculation of the molecular conductance of pipes of annular cross section, see Onusic (1980).

**Table 6.2**

Values of Factor $K_0$ in Eq. 6.11 as a Function of $d_2/d_1$[a]

| $d_2/d_1$ | 0 | 0.259 | 0.500 | 0.707 | 0.866 | 0.966 |
|---|---|---|---|---|---|---|
| $K_0$ | 1 | 1.072 | 1.154 | 1.254 | 1.430 | 1.675 |

[a]After Barret and Bosanquet (1944), quoted in Guthrie and Wackelring (1949, p. 38).

## 6.4 *Molecular Flow through Long Channels*

This type of flow is denoted by subscript m.

### 6.4.1 Molecular Conductance of Channels of Uniform Elliptical Cross Section (Dushman, 1966, p. 83)

$$C_m = (8\pi/3L)(RT/2\pi M)^{1/2}(a^2b^2)/[(a^2+b^2)/2]^{1/2} \qquad (\mathrm{cm^3\ s^{-1}}) \qquad (6.12)$$

$$= (43.1/L)(T/M)^{1/2}(a^2b^2)/(a^2+b^2)^{1/2} \qquad (\text{liters s}^{-1}) \qquad (6.12a)$$

with $R$ (ergs $\mathrm{mol^{-1}\ K^{-1}}$), $T$ (K), M (g $\mathrm{mol^{-1}}$), and $a$, $b$ (cm) the length of the semiaxes.

For air at $T$ = 23°C:

$$C_m = (138/L)(a^2b^2)/(a^2+b^2)^{1/2} \qquad (\text{liters s}^{-1}). \qquad (6.12b)$$

Steckelmacher (1978) has published calculations of the molecular conductance of pipes of elliptical cross section.

### 6.4.2 Molecular Conductance of Channels of Rectangular Cross Section

$$C_m = (8/3)(RT/2\pi M)^{1/2}a^2b^2K_1/(a+b)L \qquad (\mathrm{cm^3\ s^{-1}}) \qquad (6.13)$$

$$= 9.7(T/M)^{1/2}a^2b^2K_1/(a+b)L \qquad (\text{liters s}^{-1}) \qquad (6.13a)$$

with $R$ (ergs $\mathrm{mol^{-1}\ K^{-1}}$), $T$ (K), M (g $\mathrm{mol^{-1}}$), $a$ (cm), $b$ (cm), $b < a$ the sides of the channel (see Fig. 5.6, diagram II), $L \gg a$, $L \gg b$, and $K_1 = f(b/a)$ correction factor that accounts for the asymmetry of the cross section as compared to that of a circle. Table 6.3 lists the values of the correction factor $K_1$.

For air at $T$ = 23°C:

$$C_m = 31.0(a^2b^2)K_1/(a+b)L. \qquad (\text{liters s}^{-1}) \qquad (6.13b)$$

**Table 6.3**

Values of Factor $K_1$ in Eq. 6.13 as a Function of $b/a$[a]

| $b/a$ | 1.000 | 0.667 | 0.500 | 0.333 | 0.200 | 0.215 | 0.100 |
|---|---|---|---|---|---|---|---|
| $K_1$ | 1.108 | 1.126 | 1.151 | 1.198 | 1.297 | 1.400 | 1.444 |

[a]After Barret and Bosanquet (1944), quoted in Guthrie and Wackelring (1949, p. 39).

**Table 6.4**
Values of Factor $K_2$ in Eq. 6.14 as a Function of $L/b$[a]

| $L/b$ | 0.1 | 0.2 | 0.4 | 0.8 | 1 | 2 | 3 | 4 | 5 | 10 | $> 10$ |
|---|---|---|---|---|---|---|---|---|---|---|---|
| $K_2$ | 0.036 | 0.068 | 0.13 | 0.22 | 0.26 | 0.40 | 0.52 | 0.60 | 0.67 | 0.94 | $\ln(3/8)(L/b)$ |

[a]After Clausing, quoted in Loeb (1934).

### 6.4.3 Molecular Conductance of Slit-Like Channels

Slit-like channels have rectangular cross section of length $a$ and height $b$, such that $a \gg b$, and $b/a \ll 1$:

$$C_m = (8/3)(RT/2\pi M)^{1/2}(ab^2)K_2/L \qquad (\text{cm}^3\ \text{s}^{-1}) \qquad (6.14)$$

$$= 9.7(T/M)^{1/2}(ab^2)K_2/L \qquad (\text{liters s}^{-1}) \qquad (6.14\text{a})$$

with $R$ (ergs mol$^{-1}$ K$^{-1}$), $T$ (K), M (g mol$^{-1}$), $a$ (cm), $b$ (cm), $L$ (cm) the length of the channel, and $K_2 = f(L/b)$ correction factor that accounts for the asymmetry of the cross section as compared to that of a circle. The values of the correction factor $K_2$ are listed in Table 6.4.

For air at $T = 23°\text{C}$:

$$C_m = 31.0ab^2K_2/L \qquad (\text{liters s}^{-1}) \qquad (6.14\text{b})$$

### 6.4.4 Molecular Conductance of Channels of Triangular (Equilateral) Cross Section

$$C_m = 0.413(RT/2\pi M)^{1/2}a^3L \qquad (\text{cm}^3\ \text{s}^{-1}) \qquad (6.15)$$

$$= 1.5(T/M)^{1/2}a^3/L \qquad (\text{liters s}^{-1}) \qquad (6.15\text{a})$$

with $R$ (ergs mol$^{-1}$ K$^{-1}$), $T$ (K), M (g mol$^{-1}$), $a$ (cm) the side of the triangle, and $L$ (cm) the length of the duct. Note that the value 0.413 in Eq. 6.15 includes a correction factor $K_3 = 1.24$ that accounts for the asymmetry of the cross section as compared to that of a circle.

For air at $T = 23°\text{C}$

$$C_m = 4.8a^3/L \qquad (\text{liters s}^{-1}) \qquad (6.15\text{b})$$

## 6.5 *Molecular Flow through Short Pipes*

The following assumptions underlie at the calculation of molecular flow through short pipes (or other geometric configurations such as elbows, baffles, etc.):

The mean free path of molecules is much larger than the dimensions of the pipe.
The flow is steady-state, under conditions of $T$ = const.
The pipe provides connection between two large volumes, thus affording a random molecular entrance to the inlet plane.
The walls of the pipe are microscopically rough so that the molecules are diffusively scattered according to the cosine law.

### 6.5.1 Molecular Conductance of a Pipe of Circular Cross Section

a. *Molecular conductance calculated accounting for end effects (subscript mT).*

$$C_{mT} = C_{ma} Pr \qquad \text{(e.g., liters s}^{-1}\text{)} \tag{6.16}$$

$$= 3.64(T/M)^{1/2} A Pr \qquad \text{(liters s}^{-1}\text{)} \tag{6.16a}$$

**Table 6.5**
Values of Transmission Probability $Pr$ in Eq. 6.16 According to Different Authors

| | $Pr$ (dimensionless)[a] | | | | $Pr$ (dimensionless)[a] | | |
|---|---|---|---|---|---|---|---|
| $L/r$ | (1) | (2) | (3) | $L/r$ | (1) | (2) | (3) |
| 0.1 | 0.9639 | 0.9524 | 0.952399 | 7.0 | 0.2759 | 0.2537 | 0.247735 |
| 0.3 | 0.8989 | 0.8699 | 0.869928 | 8.0 | 0.2500 | 0.2316 | 0.225263 |
| 0.5 | 0.8421 | 0.8013 | 0.801271 | 9.0 | 0.2286 | 0.2131 | 0.206641 |
| 0.7 | 0.7921 | 0.7434 | 0.743410 | 10.0 | 0.2105 | 0.1973 | 0.190941 |
| 0.9 | 0.7477 | 0.6940 | 0.694044 | 20.0 | 0.1176 | 0.1135 | 0.109304 |
| 1.0 | 0.7273 | 0.6720 | 0.671584 | 30.0 | 0.0816 | 0.0797 | 0.076912 |
| 1.2 | 0.6897 | 0.6320 | 0.632228 | 40.0 | 0.0625 | 0.0613 | 0.059422 |
| 1.4 | 0.6557 | 0.5970 | 0.597364 | 50.0 | 0.0506 | 0.0499 | 0.048448 |
| 1.6 | 0.6250 | 0.5659 | 0.566507 | 60.0 | 0.0425 | 0.0420 | 0.040913 |
| 1.8 | 0.5970 | 0.5384 | 0.538975 | 70.0 | 0.0367 | 0.0363 | 0.035415 |
| 2.0 | 0.5714 | 0.5136 | 0.514231 | 80.0 | 0.0322 | 0.0319 | 0.031255 |
| 3.0 | 0.4706 | 0.4205 | 0.420055 | 90.0 | 0.0287 | 0.0285 | 0.027925 |
| 4.0 | 0.4000 | 0.3589 | 0.356572 | 100.0 | 0.0259 | 0.0258 | 0.025258 |
| 5.0 | 0.3478 | 0.3146 | 0.310525 | 1000.0 | 0.002659 | 0.002658 | 0.002646 |
| 6.0 | 0.3077 | 0.2807 | 0.275438 | | | | |

[a] *Sources:* column 1, from Dushman (1966, p. 91), using the relationship $Pr = 1/[1 + (3/8)(L/r)]$; from Clausing (1971, p. 636); column 3, from Cole (1976, p. 261).

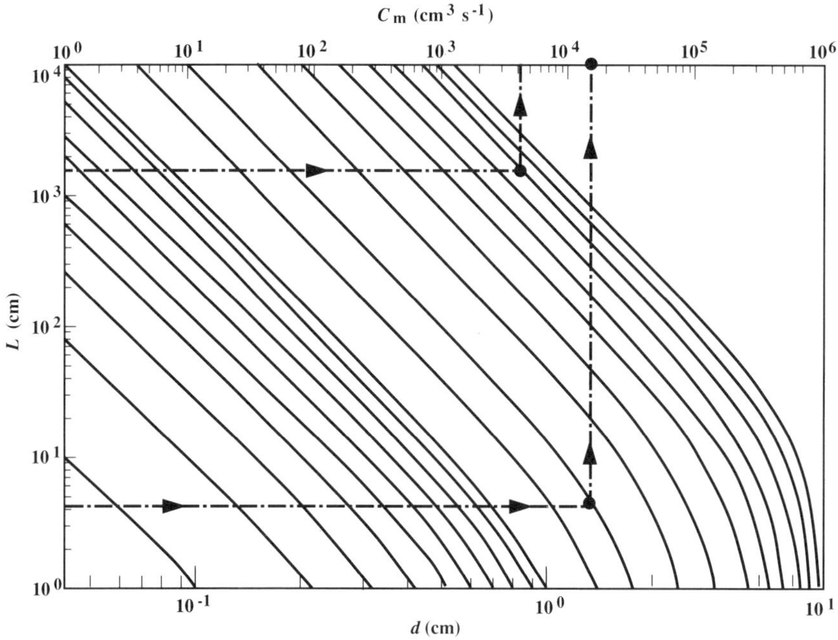


**Figure 6.4** Molecular flow conductance $C_m$ for air at $T = 23°C$ of long and short pipes of circular cross section. [After Dushman (1966).]

where $C_{ma}$ is the molecular conductance of an aperture (see Eq. 6.2c) and $Pr$ is the transmission probability that a molecule that passes through the aperture to the pipe will be transmitted to the opposite aperture.

The values of $Pr$ calculated by different authors as a function of the ratio length of the pipe ($L$) to the pipe's diameter ($d$) are listed in Table 6.5. The most accurate values of $Pr$ are those calculated by Cole (1976).

The molecular conductance for air at $T = 20°C$ of long and short pipes of circular cross section is plotted in Fig. 6.4.

*b. Molecular conductance calculated independent of end effects or transmission probabilities (subscript mp).* According to Santeler (1986)

$$C_{mp} = C_{ma}(8r/3L') \qquad \text{(e.g., liters s}^{-1}\text{)} \tag{6.17}$$

$$= 3.64(T/M)^{1/2}A(8r/3L') \qquad \text{(liters s}^{-1}\text{)} \tag{6.17a}$$

where $C_{ma}$ is the molecular conductance of the aperture to the pipe (see

Eq. 6.2c), with $T$ (K), $M$ (g mol$^{-1}$), $r$ (cm) the radius of the pipe, and $L'$ (cm) the equivalent length of the pipe.

The value of the equivalent length of the pipe $L'$ is

$$L' = L[1 + 1/(3 + 3L/7r)] \quad \text{(cm)} \tag{6.18}$$

where $L$ is the actual length of the pipe.

## 6.6 *Molecular Flow through Short Channels*

### 6.6.1 Molecular Conductance of Rectangular Slits

For a slit such as that used in molecular ray experiments, of length $a$, width $b$, and depth $L$, where $b \ll a$, $L \ll a$, $b \ll L$ (subscript ms denotes molecular conductance through slits):

$$C_{ms} = C_{ma} Pr \quad \text{(e.g., liters s}^{-1}\text{)} \tag{6.19}$$

$$= 3.64 A (T/M)^{1/2} Pr \quad \text{(liters s}^{-1}\text{)} \tag{6.19a}$$

Table 6.6 lists the values of $Pr$ in Eq. 6.19 as a function of $L/b$.

**Table 6.6**
Molecular Transmission Probability $Pr$ for Rectangular Slits in Eq. 6.19[a]

| $L/b$ | $Pr$ | $L/b$ | $Pr$ | $L/b$ | $Pr$ |
|---|---|---|---|---|---|
| 0 | 1 | 1.3 | 0.6321 | 3.2 | 0.4439 |
| 0.1 | 0.9525 | 1.4 | 0.6168 | 3.4 | 0.4318 |
| 0.2 | 0.9096 | 1.5 | 0.6024 | 3.6 | 0.4205 |
| 0.3 | 0.8710 | 1.6 | 0.5888 | 3.8 | 0.4099 |
| 0.4 | 0.8362 | 1.7 | 0.5760 | 4 | 0.3999 |
| 0.5 | 0.8048 | 1.8 | 0.5640 | 5 | 0.3582 |
| 0.6 | 0.7763 | 1.9 | 0.5525 | 6 | 0.3260 |
| 0.7 | 0.7503 | 2.0 | 0.5417 | 7 | 0.3001 |
| 0.8 | 0.7266 | 2.2 | 0.5215 | 8 | 0.2789 |
| 0.9 | 0.7049 | 2.4 | 0.5032 | 9 | 0.2610 |
| 1.0 | 0.6848 | 2.6 | 0.4865 | 10 | 0.2551 |
| 1.1 | 0.6660 | 2.8 | 0.4712 | $\infty$ | $(b/L)\ln(L/b)$ |
| 1.2 | 0.6485 | 3.0 | 0.4570 | | |

[a]From Clausing (1971, p. 642).

## *6.7 Molecular Flow through Vacuum Components of Simple and Complex Geometries*

The molecular conductance of vacuum components can be calculated by using Eq. 6.16 provided that the transmission probability $Pr$ of molecules that cross the component from both sides is known.

For a simple vacuum component, such as that illustrated in Fig. 6.5 of cross-sectional areas $A_1$ and $A_2$, connecting between infinite volumes at very low pressure the relationship

$$A_1 Pr_{1-2} = A_2 Pr_{2-1} \tag{6.20}$$

holds true when isothermal conditions are reached, since conductance is independent of direction. Here $Pr_{1-2}$ is the probability that a molecule entering the component through cross-sectional area $A_1$, would exit through cross-sectional area $A_2$; and $Pr_{2-1}$, the probability that a molecule entering the component through cross-sectional area $A_2$ would exit through cross-sectional area $A_1$.

When $A_1 = A_2$

$$Pr_{1-2} = Pr_{2-1} = Pr \qquad \text{(dimensionless)} \tag{6.21}$$

The molecular transmission probability $Pr$ depends on the geometry of the vacuum component and the cross-sectional areas to which $Pr$ is associated. Equation 6.20 holds true for the conditions mentioned in Section 6.5.

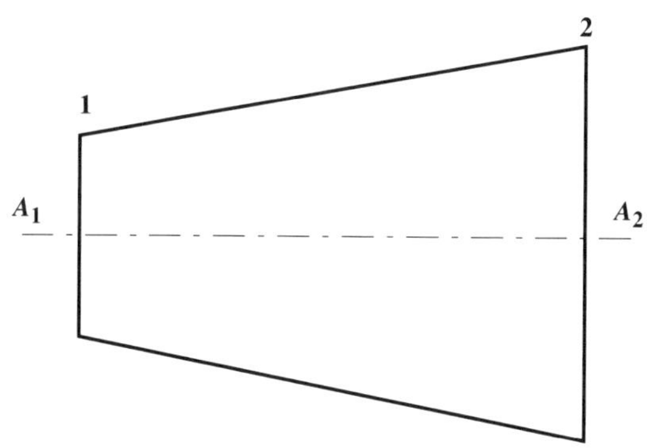


**Figure 6.5** Schematic of a vacuum component ($A_1$, $A_2$ = cross-sectional areas).

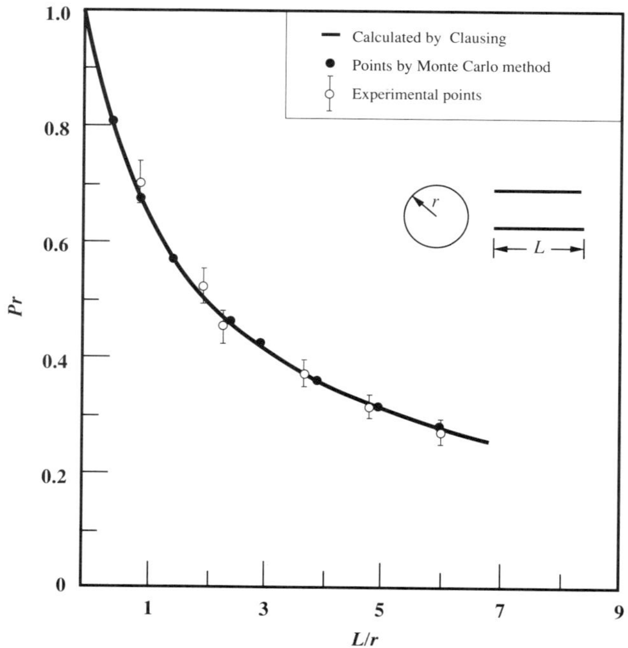


**Figure 6.6** Molecular transmission probability *Pr* for a cylindrical pipe. [After Levenson et al. (1960).]

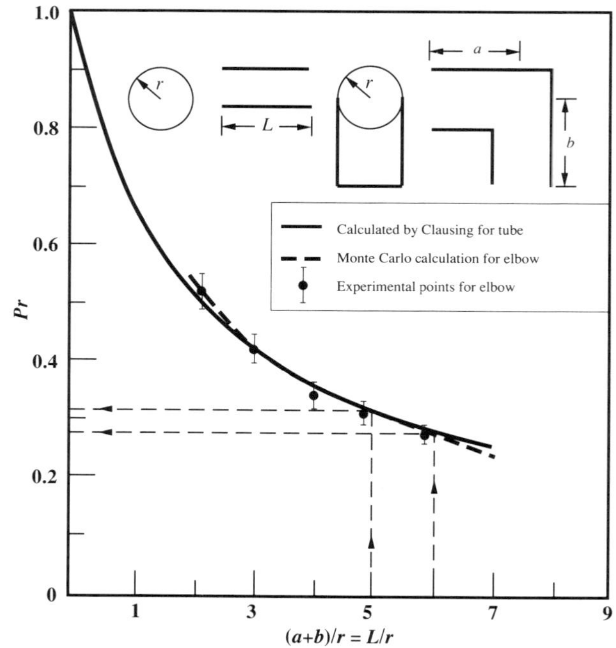


**Figure 6.7** Molecular transmission probability *Pr* for a 90° elbow. [After Levenson et al. (1960).]

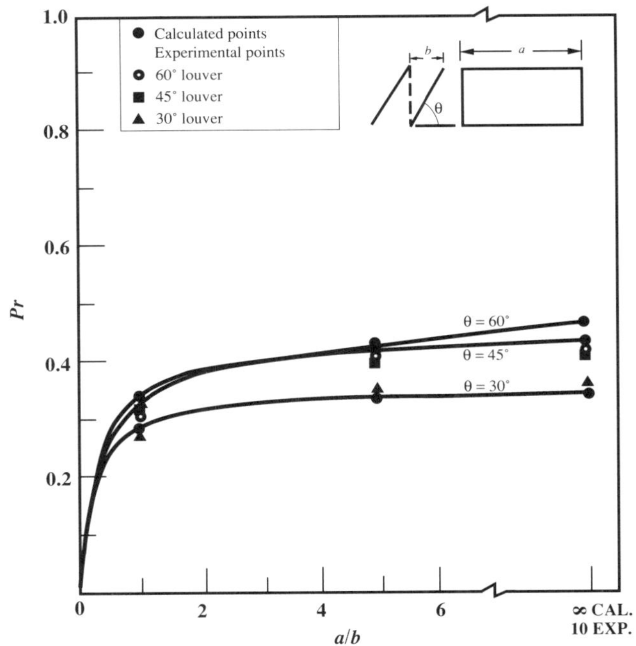


**Figure 6.8** Molecular transmission probability $Pr$ for baffles with louver geometries. [After Levenson et al. (1960).]

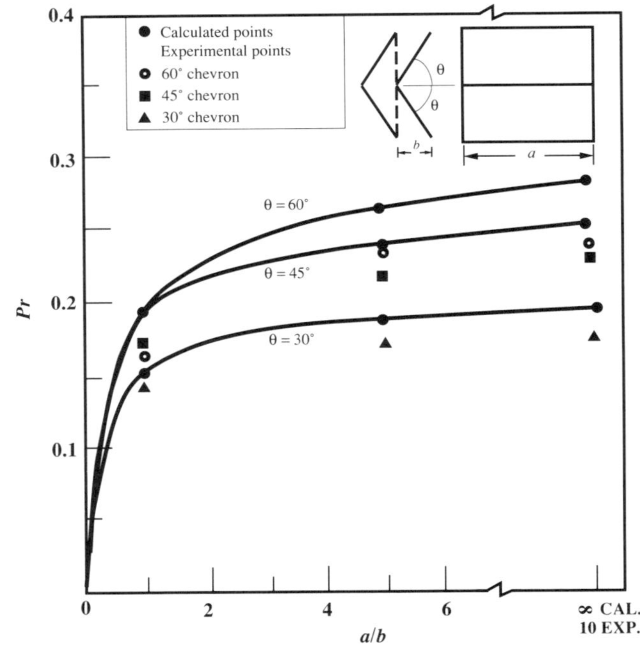


**Figure 6.9** Molecular transmission probability $Pr$ for baffles with chevron geometries. [After Levenson et al. (1960).]

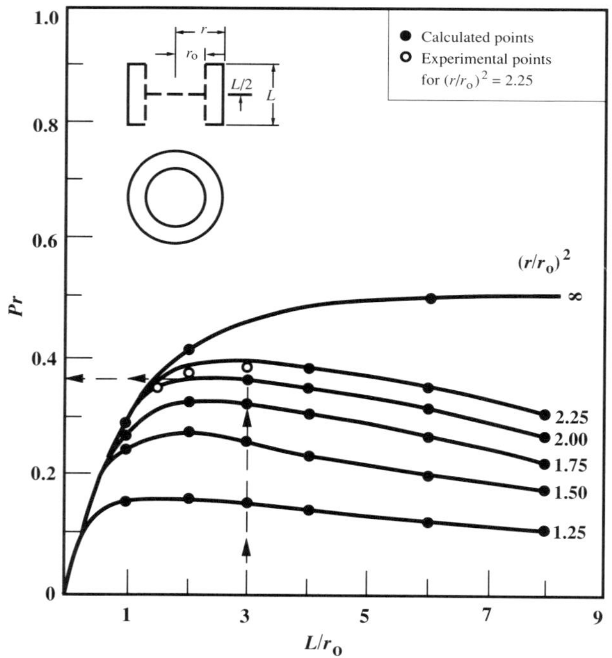


**Figure 6.10** Molecular transmission probability $Pr$ for a straight cylinder with two restricted ends and circular blocking plate. [After Levenson et al. (1960).]

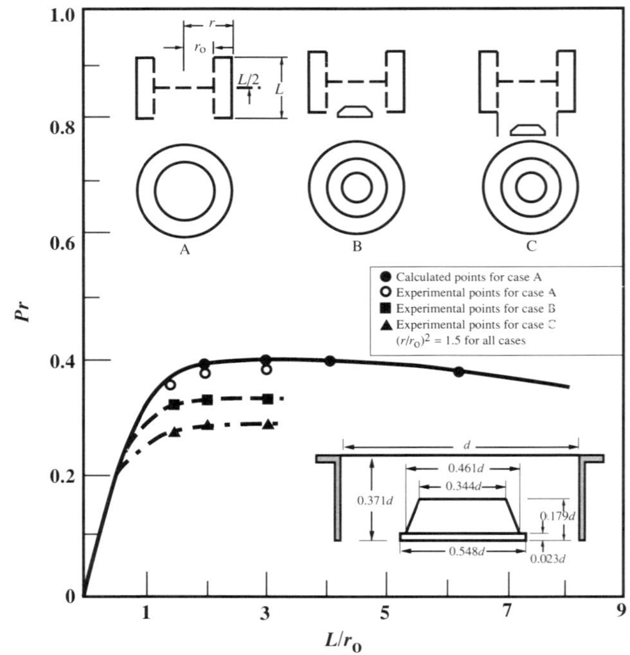


**Figure 6.11** Molecular transmission probability $Pr$ for a straight cylinder with two restricted ends and circular blocking plate in a diffusion pump system. [After Levenson et al. (1960).]

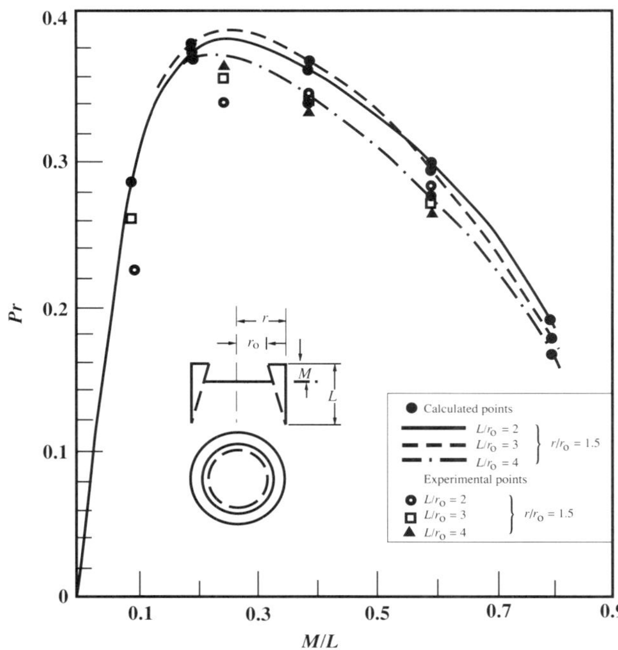


**Figure 6.12** Molecular transmission probability $Pr$ for the small end of a straight cylinder with one restricted end and circular blocking plate. [After Levenson et al. (1960).]

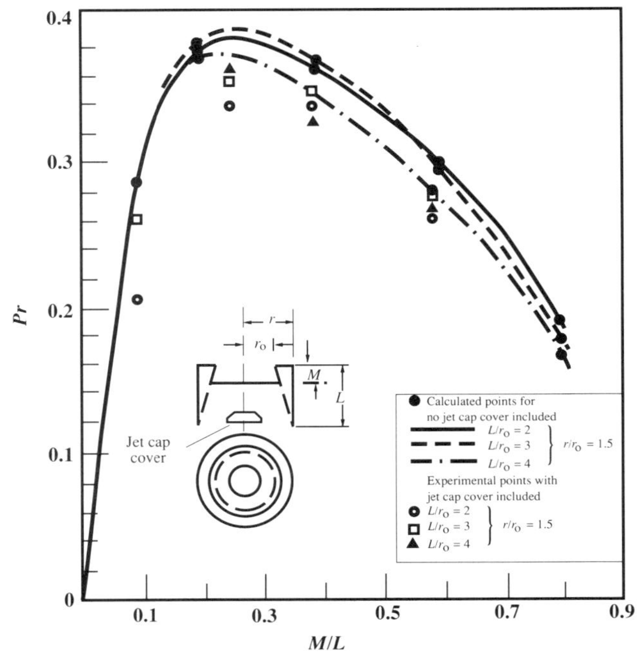


**Figure 6.13** Molecular transmission probability $Pr$ for the small end of a straight cylinder with one restricted end and circular blocking plate in a diffusion pump system. [After Levenson et al. (1963).]

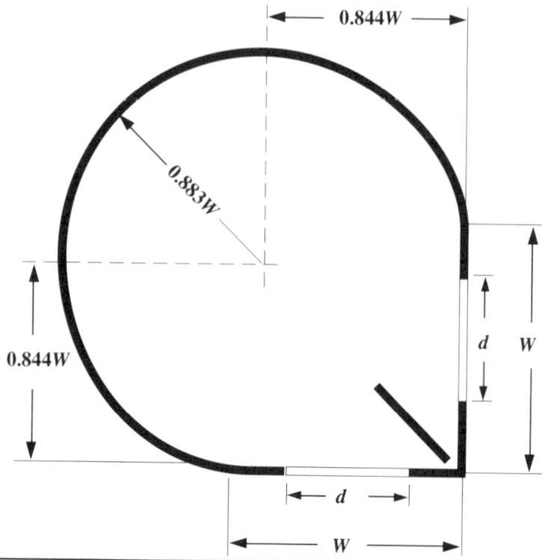


| | $W/d$ | $Pr$ Experimental |
|---|---|---|
| 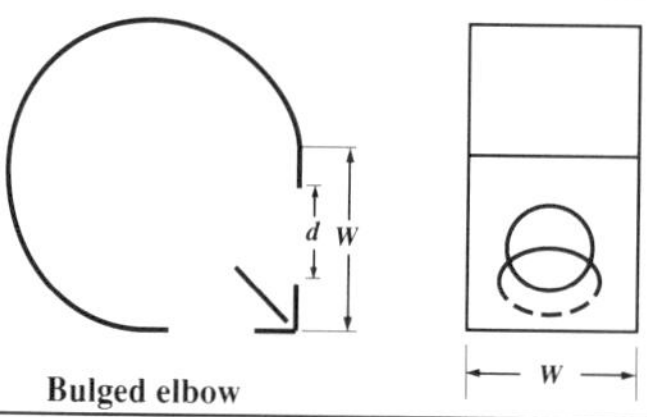  Bulged elbow | 2.00<br>1.33<br>1.00 | 0.44<br>0.39<br>0.32 |
| 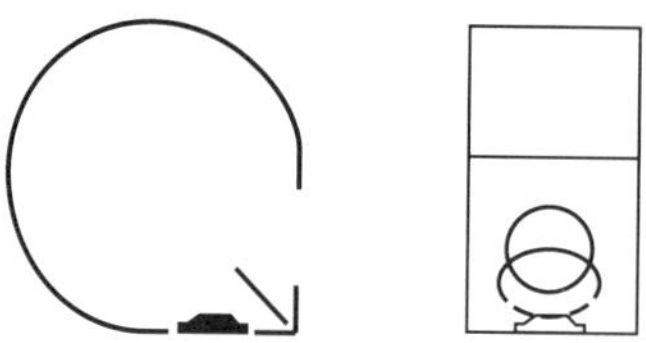 Bulged elbow with jet cap | 2.00<br>1.33 | 0.33<br>0.30 |
| 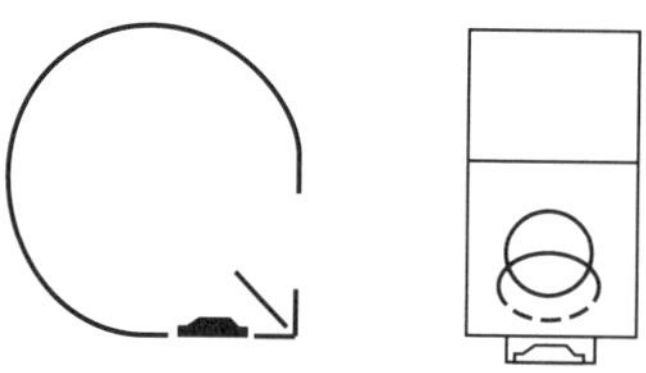 Bulged elbow on diffusion pump | 2.00<br>1.33 | 0.32<br>0.27 |
| 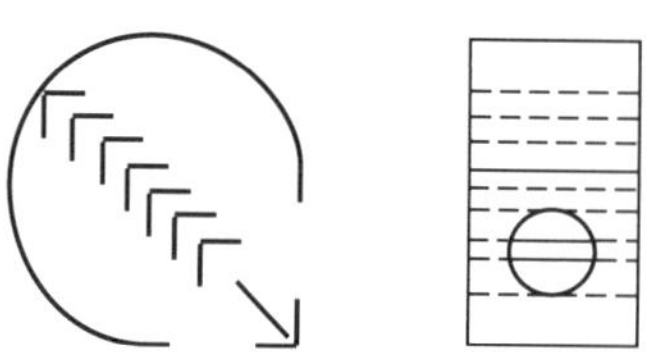 Bulged elbow with chevron $a/b$ =5 | 2.00<br>1.66<br>1.33 | 0.38<br>0.35<br>0.31 |

**Figure 6.14** Molecular transmission probability *Pr* for bulged-elbow geometries. [After Levenson et al. (1960).]

| | *W/d* | *Pr* Experimental |
|---|---|---|
| **Cubic elbow** 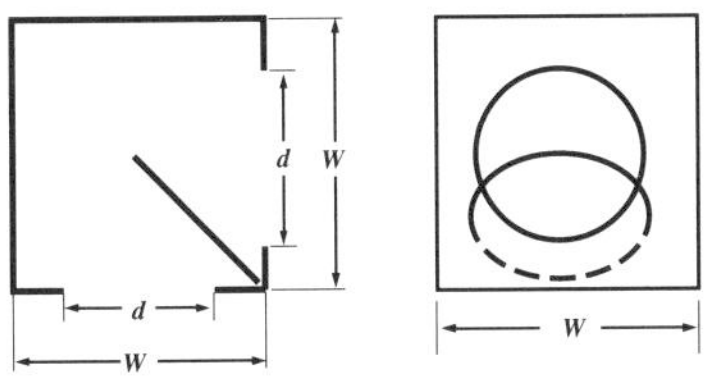  | 2.00<br>1.50 | 0.43<br>0.38 |
| **Cubic elbow with jet cap** 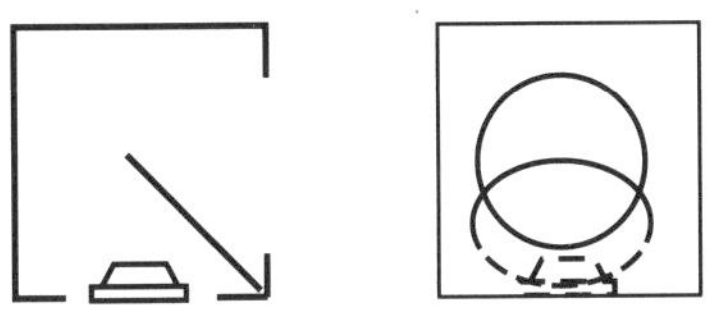 | 2.00<br>1.50 | 0.36<br>0.28 |
| **Cubic elbow on diffusion pump** 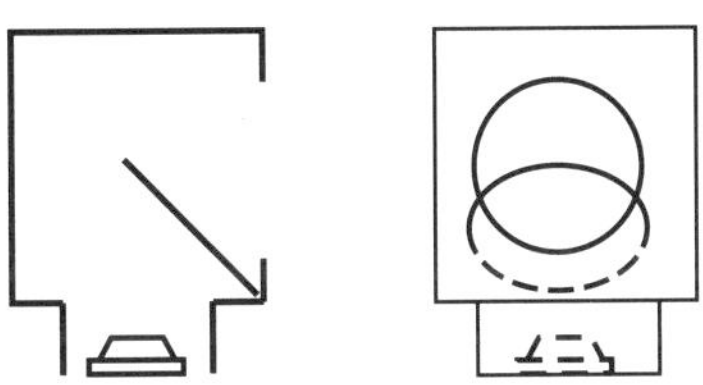 | 2.00<br>1.50 | 0.30<br>0.27 |

**Figure 6.15** Molecular transmission probability *Pr* for cubic-elbow geometries. [After Levenson et al. (1960).]

The molecular transmission probability of vacuum components calculated by using analytical and statistical methods (Monte Carlo) are illustrated in Figs. 6.6–6.18. The probable error of experimental values in Figs. 6.6–6.16 is about $\pm 6\%$ (Levenson et al., 1960).

## 6.7.1 Molecular Conductance of Vacuum Components in Series

*a. Two vacuum components of same shape, in series, matched with no discontinuity at the interface (Oatley, 1957).* For two pipes, each of cross-sectional area $A_1$

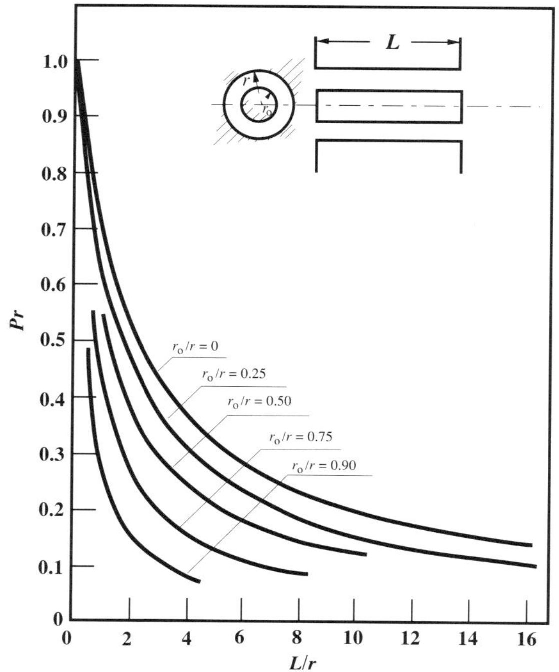


**Figure 6.16** Molecular transmission probability $Pr$ for a cylindrical annulus. [After Levenson et al. (1960).]

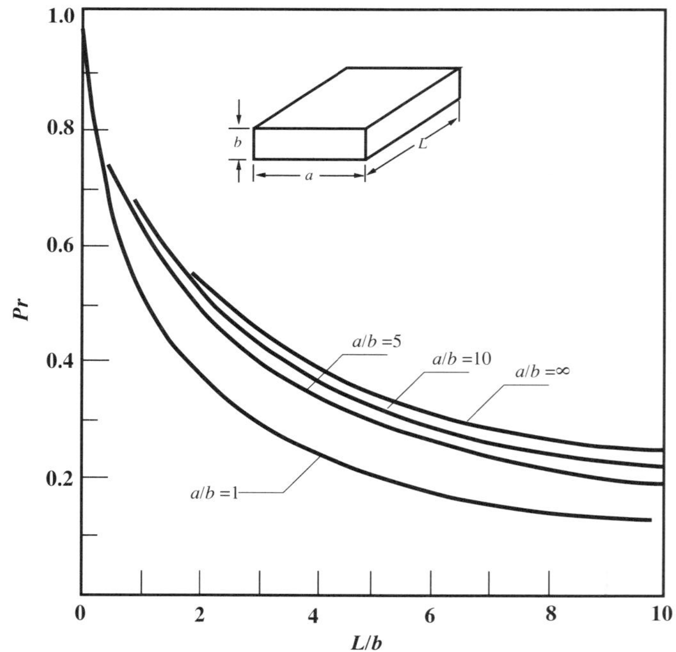


**Figure 6.17** Molecular transmission probability $Pr$ for rectangular ducts (dimensions in mm). [After Levenson et al. (1963).]

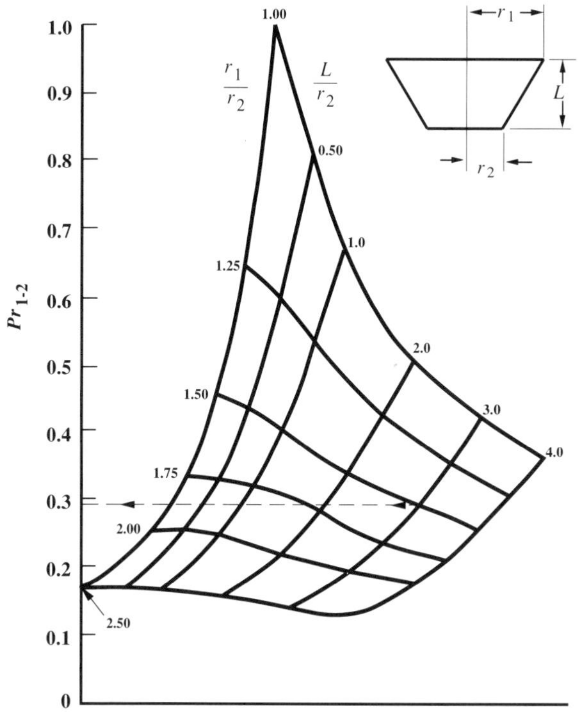


**Figure 6.18** Molecular transmission probability $Pr$ for a frustum; $Pr_{1-2}$ refers to the passage from the larger to the smaller end. [After Davis (1961).]

but of different lengths $L_1$ and $L_2$ (Fig. 6.19a):

$$1/Pr_{1-2} = 1/Pr_1 + 1/Pr_2 - 1 \qquad \text{(dimensionless)} \tag{6.22}$$

Since

$$A_1 Pr_{1-2} = A_1 Pr_{2-1} \tag{6.23}$$

$$A_1 Pr_{2-3} = A_1 Pr_{3-2} \tag{6.24}$$

$$Pr_{1-2} = Pr_{2-1} \tag{6.25}$$

$$Pr_{2-3} = Pr_{3-2} \tag{6.26}$$

$$C_{mp} = C_{ma} Pr_{1-2} \qquad \text{(e.g., liters s}^{-1}\text{)} \tag{6.27}$$

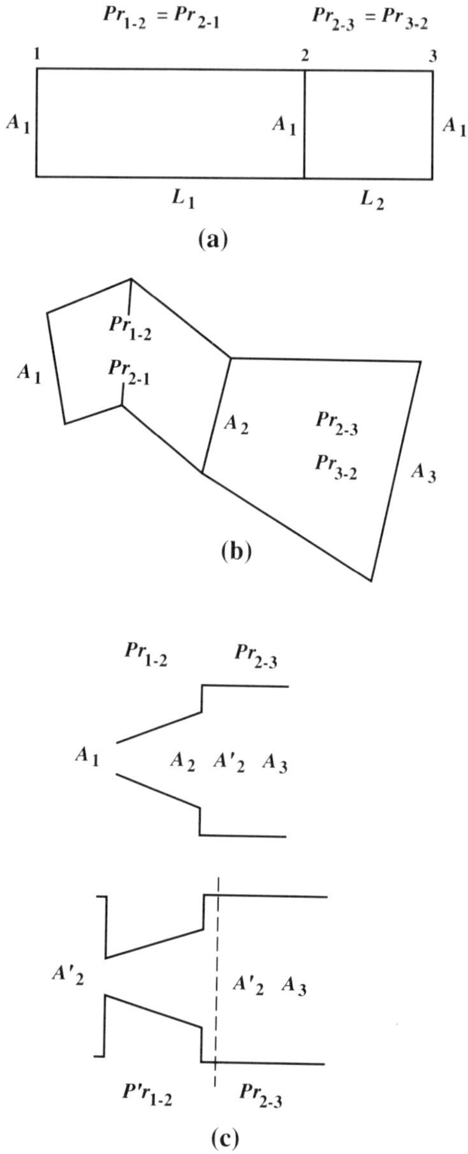


**Figure 6.19** Schematic of the method for combining the molecular transmission probabilities of two vacuum components in series: a—cylindrical pipes of same cross-sectional area $A$ and lengths $L_1$ and $L_2$; b, c—vacuum components of various shapes; $A_1$, $A_2$, $A_3$—cross-sectional areas. [After Carlson (1979).]

*b. Several vacuum components of same shape, in series, matched with no discontinuities at the interface (Oatley, 1957).*

$$(1 - Pr_{1-n})/Pr_{1-n} = (1 - Pr_1)/Pr_1 + (1 - Pr_2)/Pr_2 + \cdots + (1 - Pr_n)/Pr_n \quad \text{(dimensionless)} \quad (6.28)$$

$$C_{mp} = C_{ma} Pr_{1-n} \quad \text{(e.g., liters s}^{-1}\text{)} \quad (6.29)$$

Equations 6.22–6.25 hold true when there are not pronounced "beaming effects."

*c. Two vacuum components of different shape, in series, matched with no discontinuity at the interface (Carlson, 1979, p. 18).* Refer to Fig. 6.19b:

$$Pr_{1-3} = Pr_{1-2} Pr_{2-3}/(Pr_{2-1} + Pr_{2-3} - Pr_{2-3} Pr_{2-1}) \quad (6.30)$$

and according to Eq. 6.20

$$Pr_{1-3} = Pr_{1-2} Pr_{2-3}/[(A_1/A_2) Pr_{1-2} + Pr_{2-3} - (A_1/A_2) Pr_{1-2} Pr_{2-3}] \quad (6.31)$$

$$Pr_{3-1} = (A_1/A_3) Pr_{1-3} \quad (6.32)$$

Hence

$$C_{mp} = C_{ma_1} Pr_{1-3} \quad \text{(liters s}^{-1}\text{)} \quad (6.33)$$

*d. Two vacuum components of different shape, in series, matched with discontinuity at the interface (Carlson, 1979, p. 18).* Refer to Fig. 6.19c:

$$P'r_{1-3} = A' Pr_{1-2} Pr_{2-3}/[Pr_{1-2} + (A'/A_1) Pr_{2-3} - Pr_{1-2} Pr_{2-3}] \quad (6.34)$$

where A is the cross-sectional area at the inlet of the second vacuum component. Then

$$C_{mp} = C_{ma} Pr'_{1-3} \quad \text{(e.g., liters s}^{-1}\text{)} \quad (6.35)$$

For the transmission probability of vacuum components of more complex configurations, see Santeler and Boeckman (1987) and Santeler et al. (1966, pp. 115–120).

## *Exercises*

**Exercise 6.1**

Calculate the molecular throughput in mbar liters $s^{-1}$ and Torr liters $s^{-1}$ for air at $T = 23°C$ of a thin-edged aperture of area A.

**Solution 6.1**

Use Eq. 6.1a:

$$Q_{ma} = (1/4)(8RT/\pi M)^{1/2}(p_1 - p_2)A$$

Substitute

$R = 8.314 \times 10^7$ ergs $\text{mol}^{-1}$ $\text{K}^{-1}$ (see Table 1.3)
$T = 273 + 23 = 296$ K
$M = 28.98$ g $\text{mol}^{-1}$ (see Table 1.2)
A ($\text{cm}^2$)

When $p_1, p_2$ (dyn $\text{cm}^{-2}$)

$$Q_{ma} = (1/4)(8 \times 8.314 \times 10^7 \text{ ergs mol}^{-1} \text{ K}^{-1}$$

$$\times 296 \text{ K}/3.14 \times 28.98 \text{ g mol}^{-1})^{1/2} \times (\text{dyn cm}^{-2})(\text{cm}^2)$$

$$= 0.25 \times 46{,}514 \,(\text{ergs g}^{-1})^{1/2} \times \text{dyn} = 11{,}629 \,(\text{cm s}^{-1}) \times \text{dyn}$$

$$= 11{,}629 \,(\text{cm s}^{-1})(\text{dyn})(\text{cm}^2 \text{ cm}^{-2}) = 11{,}629 \text{ dyn cm}^{-2} \text{ cm}^3 \text{ s}^{-1}$$

When $p_1, p_2$ (mbar)

$$Q_{ma} = 11{,}629 \text{ dyn cm}^{-2} \text{ cm}^3 \text{ s}^{-1} \times 10^{-3} \text{ mbar dyn}^{-1} \text{ cm}^{-2} \times 10^{-3} \text{ liters cm}^{-3}$$

$$= 11.629 \times 10^{-3} \text{ mbar liters s}^{-1}$$

When $p_1, p_2$ (Torr)

$$Q_{ma} = 11{,}629 \text{ dyn cm}^{-2} \text{ cm}^3 \text{ s}^{-1} \times (1/1.33 \times 10^3)$$

$$(\text{Torr dyn}^{-1} \text{ cm}^{-2} 10^{-3} \text{ liters cm}^{-3})$$

$$= 8.74 \times 10^{-3} \text{ Torr liters s}^{-1}$$

**Exercise 6.2**

The pressure of a mixture of air + 1% water measured at $T = 23°\text{C}$ upstream from a thin-edged aperture is $10^{-6}$ mbar. The pressure of dry air measured downstream from the aperture is $10^{-9}$ mbar. Determine the ratio of the pumping speeds of the aperture for water vapor to air.

**Solution 6.2**

Use Eq. 6.3a:

$$S_{ma} = 3.64(T/M)^{1/2} A(1 - p_2/p_1)$$

Substitute

$$
\begin{aligned}
T &= 273 + 30 = 303 \text{ K} \\
M_{\text{air}} &= 28.98 \text{ g mol}^{-1} \text{ (see Table 1.2)} \\
M_{\text{water}} &= 18.0 \text{ g mol}^{-1} \text{ (see Table 1.2)} \\
p_{1\,\text{water}} &= 10^{-6} \text{ mbar} \times (1/100) = 10^{-8} \text{ mbar} \\
&\quad \text{(upstream from the aperture)} \\
p_{1\,\text{air}} &= 10^{-6} \text{ mbar} - 10^{-8} \text{ mbar} = 9.9 \times 10^{-7} \text{ mbar} \\
&\quad \text{(upstream from the aperture)} \\
p_{2\,\text{air}} &= 10^{-9} \text{ mbar (downstream from the aperture)}
\end{aligned}
$$

Thus

$$
\begin{aligned}
S_{\text{ma water}} &= 3.64(303/180)^{1/2} A(1 - 10^{-9}/10^{-8}) = 13.44\,A \text{ liters s}^{-1} \\
S_{\text{ma air}} &= 3.64(303/28.98)^{1/2} A(1 - 10^{-9}/9.9 \times 10^{-7}) \\
&= 11.76 \text{ liters s}^{-1} \\
S_{\text{ma water}}/S_{\text{ma air}} &= 13.44 \text{ liters s}^{-1}/11.76 \text{ liters s}^{-1} = 1.14
\end{aligned}
$$

**Exercise 6.3**

Compare the laminar and molecular conductances per unit length of long cylindrical pipes of inner diameter $d$ for air at $T = 23°\text{C}$.

**Solution 6.3**

From Eq. 5.12b the laminar conductance for air at $T = 23°\text{C}$ of a pipe of uniform cross section is

$$C_l = 138(d^4/L)\bar{p} \text{ (liters s}^{-1}\text{) when } p \text{ (mbar)}$$

From Eq. 6.9a the molecular conductance for air at $T = 23°\text{C}$ of a pipe of uniform cross section is

$$C_{\text{mp}} = 12.1(d^3/L) \qquad \text{(liters s}^{-1}\text{)}$$

Thus

$$C_l/C_{\text{mp}} = 138(d^4/L)\bar{p}/12.1(d^3/L) = 11.4\ d\bar{p}$$

with $d$ (cm) and $\bar{p}$ (mbar).

**Exercise 6.4**

Compare the molecular conductances of the aperture to a long straight pipe of circular cross section and that of the pipe itself. Comment on the results.

**Solution 6.4**

Use Eqs. 6.2b and 6.8:

$$C_{ma} = (1/4)(8RT/\pi M)^{1/2} A \qquad (\text{cm}^3\ \text{s}^{-1})$$

$$C_{mp} = (16/3)(RT/2\pi M)^{1/2}(A^2/oL) \qquad (\text{cm}^3\ \text{s}^{-1})$$

Substitute

$A = \pi r^2$
$o = 2\pi r$

Calculate $C_{ma}/C_{mp}$:

$$C_{ma}/C_{mp} = (1/4)(8RT/\pi M)^{1/2}$$

$$\times\ \pi r^2/(16/3)(RT/2\pi M)^{1/2}(\pi^2 r^4/2\pi rL)$$

$$= (3/8)(L/r)$$

*Comments*: The impedance of the aperture is $Z_{ma} = 1/C_{ma}$, whereas the impedance of the pipe itself is $Z_{mp} = 1/C_{mp}$. The ratio $Z_{mp}/Z_{ma}$ is

$$Z_{mp}/Z_{ma} = (3/8)(L/r)$$

Assume first, that the length of the pipe $L$ is 40 times larger than the radius $r$ of the aperture to the pipe. Thus

$$Z_{mp}/Z_{ma} = (3/8) \times 40 = 15$$

For such a pipe, the molecular impedance of the pipe itself is 15 times higher than the molecular impedance of the aperture.

If one considers a pipe whose length $L$ is 10 times larger than the radius of the aperture to the pipe (short pipe), then

$$Z_{mp}/Z_{ma} = (3/8) \times 10 = 3.75$$

or

$$C_{ma}/C_{mp} = 3.75$$

which gives

$$C_{mp} = C_{ma}/3.75 = 0.27C_{ma}$$

which obviously is erroneous.

**Exercise 6.5**

Estimate the conductances of an aperture of cross-sectional area $A$ for the gases $H_2$, He, air, $CO_2$, Kr, and Xe, all at $T = 23°\text{C}$.

**Solution 6.5**

Use Eq. 6.2c:

$$C_{ma} = 3.64(T/M)^{1/2} A$$

Substitute

$T = 273 + 23 = 296$ K
$M_{H_2} = 2.0158$ g mol$^{-1}$ (see Table 1.2)
$M_{He} = 4.0026$ g mol$^{-1}$ "
$M_{air} = 28.98$ g mol$^{-1}$ "
$M_{CO_2} = 44.01$ g mol$^{-1}$ "
$M_{Kr} = 83.80$ g mol$^{-1}$ "
$M_{Xe} = 131.30$ g mol$^{-1}$ "

Thus

$$C_{ma\,H_2} = 3.64(296/2.0158)^{1/2}A = 12.12A \quad \text{(liters s}^{-1}\text{)}$$

$$C_{ma\,He} = 3.64(296/4.0026)^{1/2}A = 31.30A \quad "$$

$$C_{ma\,air} = 3.64(296/28.98)^{1/2}A = 11.63A \quad "$$

$$C_{ma\,CO_2} = 3.64(296/44.01)^{1/2}A = 9.44A \quad "$$

$$C_{ma\,Kr} = 3.64(296/83.80)^{1/2}A = 6.84A \quad "$$

$$C_{ma\,Xe} = 3.64(296/131.30)^{1/2}A = 5.47A \quad "$$

The molecular conductance of an aperture varies inversely as $(1/M)^{1/2}$. This property can be used to separate gases of different species.

**Exercise 6.6**

What is the molecular conductance for air and argon at $T = 23°C$ of a pipe 1 m long and of 2 cm inner diameter?

**Solution 6.6**

Since $L(100 \text{ cm}) > 20 \times 2$ cm, Eq. 6.9 applies:

$$C_{mp} = 3.81(T/M)^{1/2} d^3/L$$

Substitute

$T = 273 + 23 = 296$ K
$M_{air} = 28.98$ g mol$^{-1}$ (see Table 1.2)
$M_{Ar} = 39.948$ g mol$^{-1}$ "
$d = 2$ cm
$L = 100$ cm

Hence

$$C_{mp\ air} = 3.81(296/28.98)^{1/2} \times 2^3/100 = 0.974 \text{ liters s}^{-1}$$

$$C_{mp\ Ar} = 3.81(296/39.948)^{1/2} \times 2^3/100 = 0.830 \text{ liters s}^{-1}$$

Note that if the molecular conductance of a pipe for air at a certain temperature is known, the molecular conductance of that pipe for other gases, at the same temperature, is easily obtained using the relation

$$C_{mp\ gas}/C_{mp\ air} = (M_{air}/M_{gas})^{1/2}$$

Thus

$$C_{mp\ Ar} = (28.98/39.948)^{1/2} \times 0.974 = 0.830 \text{ liters s}^{-1}$$

The molecular conductance of a pipe varies as $(1/M)^{1/2}$.

**Exercise 6.7**

A turbomolecular pump of capacity 50 liters s$^{-1}$ pumps air at $T = 23$°C from a vessel, through a tapered pipe 190 cm long. The diameters at the two ends of the pipe are 7 cm and 11 cm, respectively. Calculate the net pumping speed at the port of the vessel. Refer to Fig. 6A.

From Eqs. 4.41 and 6.10a

$$S_n = S_P C/(S_p + C)$$

$$C_{mp} = 24.35(d_1^2 d_2^2)/[(d_1 + d_2)L]$$

Substitute

$d_1 = 7$ cm
$d_2 = 11$ cm
$L = 190$ cm
$S_P = 50$ liters s$^{-1}$

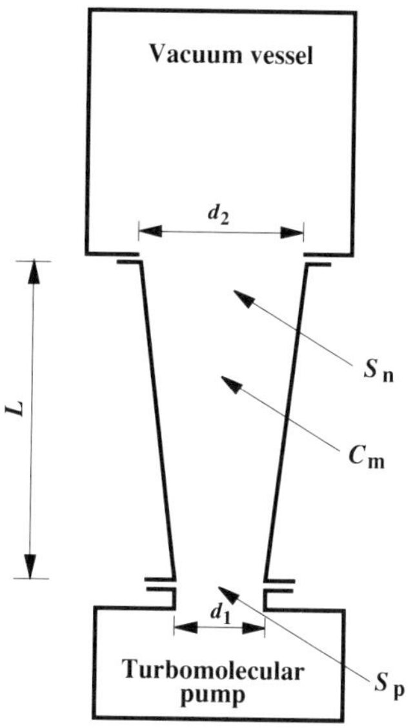


Figure 6A

Calculate

$$C_{mp} = 24.35\ (7^2 \times 11^2)/(7 + 11) \times 190 = 42.2 \text{ liters s}^{-1}$$

$$S_n = 50 \text{ liters s}^{-1} \times 42.2 \text{ liters s}^{-1}/(50 \text{ liters s}^{-1} + 42.2 \text{ liters s}^{-1})$$

$$= 22.89 \text{ liters s}^{-1}$$

**Exercise 6.8**

Use the graph in Fig. 6.3 to evaluate the molecular conductance of a pipe 150 cm long of 3.81 cm diameter fitting a con-flat flange (CFF) measuring 2.75 in. Check the result.

**Solution 6.8**

The vertical interrupted line passing through $L$ = 150 cm intercepts the line marked "CFF 2.75″ [inches]" at $C_{mp}$ = 4.5 liters $s^{-1}$. Check the result using Eq. 6.9a:

$$C_{mp} = 12.1d^3/L$$

Substitute

$d$ = 3.81 cm
$L$ = 150 cm

Thus

$$C_{mp} = 12.1 \times 3.81^3/150 = 4.46 \text{ liters s}^{-1}$$

**Exercise 6.9**

Calculate the molecular conductance of a slit-like channel 0.05 cm long and with sides 5 cm and 0.005 cm, for Ar at $T = 0°C$.

**Solution 6.9**

The dimensions of the slit-like channel are

$L = 0.05$ cm
$a = 5$ cm (refer to Fig. 5.6, II)
$b = 0.005$ cm

Check whether $a \gg b$ and $b/a \ll 1$ (see Section 6.4.3). For this case

$$5 \text{ cm} \gg 0.005 \text{ cm}$$
$$0.005 \text{ cm}/5 \text{ cm} \ll 1$$

The channel is long and Eq. 6.14a applies:

$$C_{mp} = 9.7(T/M)^{1/2} ab^2 K_2/L$$

Substitute

$T = 273 + 0 = 273$ K
$M_{Ar} = 39.948$ g mol$^{-1}$ (see Table 1.2)

for $L/b = 0.05/0.005 = 10$, $K_2 = 0.49$ (see Table 6.4). Thus

$$C_m = 9.7(273/39.948)^{1/2} \times 5 \times 0.005^2 \times 0.94/0.05$$
$$= 5.96 \times 10^{-2} \text{ liters s}^{-1}$$

**Exercise 6.10**

Calculate the molecular conductance of a long pipe of (a) elliptical cross section, (b) rectangular cross section, and (c) triangular (equilateral) cross section, assuming that the molecular conductance of the pipe is of uniform cross section.

**Solution 6.10**

Use Eq. 6.8:

$$C_{mp} = (16/3)(RT/2\pi M)^{1/2} A^2/oL$$

(a) The molecular conductance of a pipe of elliptical cross section. If $a$ and $b$ are respectively the semimajor and semiminor axes of the elliptical cross section, then

$$A = \pi ab \qquad (\mathrm{cm}^2)$$

$$o = 2\pi[(a^2 + b^2)/2]^{1/2} \qquad (\mathrm{cm})$$

Hence

$$C_{\mathrm{mp}} = (16/3)(RT/2\pi M)^{1/2} \times \pi^2a^2b^2/2\pi L[(a^2 + b^2)/2]^{1/2}$$

Rearranging the terms, we obtain

$$C_{\mathrm{mp}} = (8\pi/3L)(RT/2\pi M)^{1/2}(a^2b^2)/[(a^2 + b^2)/2]^{1/2} \qquad (\mathrm{cm}^3\ \mathrm{s}^{-1})$$

which is Eq. 6.12.

(b) Molecular conductance of a pipe of rectangular cross section. If $a$ and $b$ are the sides of the rectangle, then

$$A = ab \qquad (\mathrm{cm}^2)$$

$$o = 2(a + b) \qquad (\mathrm{cm})$$

Substitute $A$ and $o$ in Eq. 6.8. Thus

$$C_{\mathrm{mp}} = (16/3)(RT/2\pi M)^{1/2}a^2b^2/2(a + b)L$$

$$= (8/3)(RT/2\pi M)^{1/2}a^2b^2K_1/(1 + b)L \qquad (\mathrm{cm}^3\ \mathrm{s}^{-1})$$

where $K_1$ is a factor that accounts for the asymmetry of the rectangular cross section as compared to that of a circle. Thus Eq. 6.13 was obtained.

(c) Molecular conductance of a pipe of triangular (equilateral) cross section. The cross-sectional area $A$ and the perimeter $o$ of an equilateral triangle of sides $a$ are

$$A = a^2(3)^{1/2}/4 \qquad (\mathrm{cm}^2)$$

$$o = 3a \qquad (\mathrm{cm})$$

Substitute these values in Eq. 6.8. Thus

$$C_{\mathrm{mp}} = (16/3)(RT/2\pi M)^{1/2} \times a^4 \times 3/16/3aL \qquad (\mathrm{cm}^3\ \mathrm{s}^{-1})$$

$$= 0.3333(RT/2\pi M)^{1/2}a^3/L$$

Using a correction factor $K_3$ to account for the asymmetry of the triangular (equilateral) cross section compared to that of a circle, we obtain

$$C_{mp} = 0.3333 \times 1.24(RT/2\pi M)^{1/2} a^3 L$$

$$= 0.413(RT/2\pi M)^{1/2} a^3 L$$

which is Eq. 6.15.

**Exercise 6.11**

Compare the molecular conductances for air at $T = 23°C$ of long straight pipes of (a) uniform circular cross section and of channels of (b) square cross section and, (c) triangular (equilateral) cross section.

**Solution 6.11**

Assume that the pipe (a) and the channels [(b) and (c)] have the same length $L$ and that the sides of the channels are equal to the diameter of the pipe (refer to Fig. 5.9). This makes it possible to compare their molecular conductance.

Use Eqs. 6.9a, 6.13b, and 6.15b and subscripts □ and △ to specify the shape of the cross section.

(a) Pipe of uniform circular cross section:

$$C_{mp} = 12.1d^3/L \qquad (\text{liters s}^{-1})$$

with $d$ (cm) and $L$ (cm).

(b) Duct of square cross section. Write Eq. 6.13b for $a = b = d$

$$C_{m\square} = 31d^4 K_1/2dL \qquad (\text{liters s}^{-1})$$

and substitute the value of $K_1 = 1.108$ (Table 6.3) for $b/a = 1$. Thus

$$C_{m\square} = 17.174d^3/L \qquad (\text{liters s}^{-1})$$

(c) Duct of triangular (equilateral) cross section:

$$C_{m\triangle} = 4.8d^3/L \qquad (\text{liters s}^{-1})$$

Now

$$C_{m\square}/C_{mp} = (17/174d^3/L)\text{ liters s}^{-1}/(12.1d^3/L)\text{ liters }^{-1} = 1.419$$

$$C_{m\triangle}/C_{mp} = (4.8d^3/L)\text{ liters s}^{-1}/(12.1d^3/L)\text{ liters s}^{-1} = 0.397$$

**Exercise 6.12**

Calculate the total molecular conductance of a pipe using the sum of reciprocal molecular conductances of the aperture to the pipe and the conductance of a pipe of uniform cross section.

**Solution 6.12**

Write Eq. 4.33 as

$$1/C_{mT} = 1/C_{ma} + 1/C_{mp}$$

where $C_{mT}$ is the total molecular conductance of the pipe, $C_{ma}$ is the molecular conductance of the aperture to the pipe, and $C_{mp}$ represents the molecular conductance of the pipe itself.

Substitute $C_{ma}$ and $C_{mp}$ using Eqs. 6.2b and 6.8:

$$C_{ma} = (1/4)(8RT/\pi M)^{1/2} A$$

$$C_{mp} = (16/3)(RT/2\pi M)^{1/2} A^2/oL$$

$$= (4/3)(8RT/\pi M)^{1/2} A^2/oL$$

Thus

$$C_{mT} = [(1/4)(8RT/\pi M)^{1/2} A \times (4/3)(8RT/\pi M)^{1/2}(A^2/oL)]/$$

$$[(1/4)(8RT/\pi M)^{1/2} A + (4/3)(8RT/\pi M)^{1/2}(A^2/oL)]$$

$$= [(1/3)(8RT/\pi M)^{1/2} A \times (A/oL)]/(1/4 + 4/3 \times A/oL)$$

$$= (1/3)(8RT/\pi M)^{1/2} A \times 12A/(3oL + 16A)$$

$$= (1/3)(8RT/\pi M)^{1/2} A \times 12/(13oL/A + 16)$$

Substitute

$$o/A = 2\pi r/\pi r^2 = 2/r$$

Hence

$$C_{mT} = (1/3)(8RT/\pi M)^{1/2} A \times 6/(3L/r + 8)$$

$$= (1/3)(RT/\pi M)^{1/2} A \times [(6/8)(1 + (3/8)(L/r)]$$

$$= (1/4)(8RT/\pi M)^{1/2} A \times 1/[1 + (3/8)(L/r)]$$

noting that

$$Pr = 1/[1 + (3/8)(L/r)]$$
$$C_{ma} = (1/4)(8RT/\pi M)^{1/2} A$$
$$C_{mT} = C_{ma} Pr$$

The expression $1/[1 + (3/8)(L/r)]$ approximates the total conductance of a cylindrical pipe (Dushman, 1966, p. 91). Dushman calls $C_{ma}$ the *end effect*.

**Exercise 6.13**

Use the graph in Fig. 6.4 to evaluate the conductance for air at $T = 20°C$ of a pipe: (a) 8 cm inner diameter, 1400 cm long; (b) 2 cm inner diameter, 4.3 cm long. Check the results.

**Solution 6.13**

The graph in Fig. 6.4 permits the evaluation of the molecular conductance of long pipes on the linear portion of the plot and of short pipes on the curved portion of the plot. The pipe specified under (a) is long, whereas that specified under (b) is short.

(a) The interrupted line passing through the ordinate $L = 1400$ cm intersects on the graph the linear region of the curve $d = 8$ cm, at $C_m = 4.3 \times 10^3$ cm$^3$ s$^{-1}$ = 4.3 liters s$^{-1}$. To check this result, use Eq. 6.9:

$$C_{mp} = 3.81(T/M)^{1/2}d^3/L$$

Substitute

$$\begin{aligned} T &= 273 + 20 = 293 \text{ K} \\ M &= 28.98 \text{ g mol}^{-1} \text{ (see Table 1.2)} \\ d &= 8 \text{ cm} \\ L &= 1400 \text{ cm} \\ C_{mp} &= 3.81(293/28.98)^{1/2} \times 8^3/1400 = 4.43 \text{ liters s}^{-1} \end{aligned}$$

(b) The interrupted line passing through the ordinate $L = 4.3$ cm intersects on the graph the curved part of the curve $d = 2$ cm at $C_m = 1.4 \times 10^4$ cm$^3$ s$^{-1}$ = 14 liters s$^{-1}$. To check this result, use Eq. 6.16a:

$$C_{mT} = 3.64(T/M)^{1/2}APr$$

where, according to Dushman (1966, p. 91)

$$Pr = 1/[1 + (3/8)(L/r)]$$

Substitute

$$\begin{aligned} T &= 273 + 20 = 293 \text{ K} \\ M &= 28.98 \text{ g mol}^{-1} \text{ (see Table 1.2)} \\ d &= 2 \text{ cm} \\ A &= \pi d^2/4 = 3.14 \\ L &= 4.3 \text{ cm} \\ Pr &= 1/[1 + (3/8)(4.3/1)] = 0.38278 \text{ (see also Table 6.5)} \end{aligned}$$

Thus

$$C_{mT} = 3.64(293/28.98)^{1/2} \times 3.14 \times 0.38278 = 13.91 \text{ liters s}^{-1}$$

**Exercise 6.14**

Calculate the molecular conductance for air at $T = 23°C$ of a short pipe, 75 cm long of 5 cm diameter, that connects between two large volumes at very low pressure. Address both the transmission probabilities according to Dushman (1966) and Clausing (1932). Comment on the results.

**Solution 6.14**

Use Eq. 6.16a:

$$C_{mT} = 3.64(T/M)^{1/2} APr$$

Substitute

$T = 273 + 23 = 296$ K
$M = 28.98$ g mol$^{-1}$ (see Table 1.2)
$d = 5$ cm, $r = 2.5$ cm
$A = \pi d^2/4 = 3.14 \times 5^2$ cm$^2/4 = 19.625$ cm$^2$
$L = 7.5$ cm

For the ratio $L/r = 7.5$ cm$/2.5$ cm $= 3$, the molecular transmission probability (see Table 6.5) is

$Pr = 0.4706$ according to Dushman
$Pr = 0.4205$ according to Clausing

Thus, according to Dushman

$$C_{mT} = 3.64(296/28.98)^{1/2} \times 19.625 \times 0.4706 = 107.438 \text{ liters s}^{-1}$$

and according to Clausing

$$C_{mT} = 3.64(296/28.98)^{1/2} \times 19.625 \times 0.4205 = 96.000 \text{ liters s}^{-1}$$

Calculations of the molecular conductivity of short pipes (Eq. 6.16a) using the molecular transmission probability according to Clausing are more accurate than are those using the molecular transmission probability according to Dushman. The error resulting in calculations is

$$C_{mT,\,Dushman}/C_{mT,\,Clausing} = 100 \times [(107.438 - 96.000)/(96.000)] = 11.91\%$$

It can be seen in the following table that the error made in calculations using Dushman's molecular transmission probability amounts to 11.59% for ratios $L/r \leqslant 3$. Then the error is lower, becoming 0.037% for $L/r = 1000.0$.

| | *Pr* | | % | | *Pr* | | % |
|---|---|---|---|---|---|---|---|
| $L/r$ | Dushman[a] | Clausing[b] | error | $L/r$ | Dushman[a] | Clausing[b] | error |
| 0.1 | 0.9639 | 0.9524 | 1.20 | 7.0 | 0.2759 | 0.2537 | 8.75 |
| 0.3 | 0.8989 | 0.8699 | 3.33 | 8.0 | 0.2500 | 0.2316 | 7.94 |
| 0.5 | 0.8421 | 0.8013 | 5.09 | 9.0 | 0.2286 | 0.2131 | 7.27 |
| 0.7 | 0.7921 | 0.7434 | 6.55 | 10.0 | 0.2105 | 0.1973 | 6.99 |
| 0.9 | 0.7477 | 0.6940 | 7.74 | 20.0 | 0.1176 | 0.1135 | 3.61 |
| 1.0 | 0.7273 | 0.6720 | 8.23 | 30.0 | 0.0816 | 0.0797 | 2.38 |
| 1.2 | 0.6897 | 0.6320 | 9.13 | 40.0 | 0.0625 | 0.0613 | 1.96 |
| 1.4 | 0.6557 | 0.5970 | 9.83 | 50.0 | 0.0506 | 0.0499 | 1.40 |
| 1.6 | 0.6250 | 0.5659 | 10.44 | 60.0 | 0.0425 | 0.0420 | 1.19 |
| 1.8 | 0.5970 | 0.5384 | 10.88 | 70.0 | 0.0367 | 0.0363 | 1.10 |
| 2.0 | 0.5714 | 0.5136 | 11.25 | 80.0 | 0.0322 | 0.0319 | 0.94 |
| 3.0 | 0.4706 | 0.4205 | 11.91 | 90.0 | 0.0287 | 0.0285 | 0.70 |
| 4.0 | 0.4000 | 0.3589 | 11.45 | 100.0 | 0.0259 | 0.0258 | 0.38 |
| 5.0 | 0.3478 | 0.3146 | 10.55 | 1000.0 | 0.002659 | 0.002658 | 0.037 |
| 6.0 | 0.3077 | 0.2807 | 9.62 | | | | |

[a]Calculated from $Pr = 1/[1 + (3/8)(L/r)]$ (see Exercise 6.12).
[b]From Clausing (1971, p. 641).

**Exercise 6.15**

Calculate the molecular conductance for air at $T = 23°C$ of a pipe 12.6 cm long, of 6.3 cm inner diameter, independent of the end effects or transmission probability. Assume a random gas entrance and exit.

**Solution 6.15**

From Eqs. 6.17a and 6.18:

$$C_{mp} = 3.64(T/M)^{1/2} A(8r/3L')$$
$$L' = L[1 + 1/(3 + 3L/7r)]$$

Substitute

$T = 273 + 23 = 296$ K
$M = 28.98$ g mol$^{-1}$ (see Table 1.2)
$d = 6.3$ cm, $r = 3.15$ cm, $A = \pi r^2 = 31.16$ cm$^2$
$L = 12.6$ cm

Calculate the equivalent length $L'$ of the pipe:

$$L' = 12.6 \text{ cm} \left[1 + 1/(3 + 3 \times 12.6 \text{ cm}/7 \times 3.15 \text{ cm})\right] = 15.27 \text{ cm}$$

Thus

$$C_{mp} = 3.64(296/28.98)^{1/2} \times 31.16 \times (8 \times 3.15/3 \times 15.27)$$
$$= 199.40 \text{ liters s}^{-1}$$

**Exercise 6.16**

A turbomolecular pump of 1400 liters $s^{-1}$ at the working point is connected to a large vessel by a pipe 21.3 cm long and of 10.65 cm inner radius. The gas pumped is air at $T = 23°C$. Assuming a random distribution of air at the inlet and outlet of the pipe, calculate the net pumping speed at the port of the vessel using the molecular conductance of the pipe independent of any end effects and transmission probability. What is the resulting error in calculating the net pumping speed at the port of the vessel, using the conventional approach for calculating the molecular conductance of the pipe, without subtracting as reciprocals the molecular conductance of the aperture to the pipe.

**Solution 6.16**

From Eq. 4.41, the net pumping speed $S_n$ at the port of the vessel is

$$S_n = S_P C_{mp}/(S_P + C_{mp})$$

The molecular conductance $C_{mp}$ of the pipe, independent of any end effects or transmission probability (Eqs. 6.17 and 6.18), is

$$C_{mp} = 3.64(T/M)^{1/2} A(8r/3L')$$

where

$$L' = L[1 + 1/(3 + 3L/7r)]$$

Substitute

$T = 273 + 23 = 296$ K
$M = 28.98$ g $mol^{-1}$ (see Table 1.2)
$r = 10.65$ cm
$A = \pi r^2 = 3.14 \times 10.65^2\ cm^2 = 356.15\ cm^2$
$L = 21.3$ cm
$S_P = 1400$ liters $s^{-1}$

Thus

$$L' = 21.3\ \text{cm}\ \{1 + 1/[3 + (3 \times 21.3\ \text{cm}/7 \times 10.65\ \text{cm})]\} = 26.82\ \text{cm}$$

$$C_{mp} = 3.64 \times (296/28.98)^{1/2} \times 356.15 \times 8 \times 10.65/(3 \times 26.82)$$

$$= 4387\ \text{liters s}^{-1}$$

and the net pumping speed at the port of the vessel is

$$S_n = 1400\ \text{liters s}^{-1} \times 4387\ \text{liters s}^{-1}/(1400\ \text{liters s}^{-1} + 4387\ \text{liters s}^{-1})$$

$$= 1061\ \text{liters s}^{-1}$$

Calculate again the net pumping speed at the port of the vessel using the molecular conductance of the pipe given by Eq. 6.16a

$$C_{mT} = 3.64(T/M)^{1/2} APr$$

and substitute

$$\begin{aligned} T &= 273 + 23 = 296 \text{ K} \\ M &= 28.98 \text{ g mol}^{-1} \\ L &= 21.3 \text{ cm} \\ r &= 10.65 \text{ cm} \\ A &= \pi r^2 = 3.14 \times 10.65^2 \text{ cm}^2 = 356.15 \text{ cm}^2 \\ L/r &= 21.3 \text{ cm}/10.65 \text{ cm} = 2 \end{aligned}$$

Then $Pr = 0.514231$ according to Cole (1976) in Table 6.5. Thus

$$C_{mT} = 3.64(296/28.98)^{1/2} \times 356.15 \times 0.514231 = 2130.53 \text{ liters s}^{-1}$$

and

$$\begin{aligned} S_n &= 1400 \text{ liters s}^{-1} \times 2130.53 \text{ liters s}^{-1} / \\ &\quad (1400 \text{ liters s}^{-1} + 2130.53 \text{ liters s}^{-1}) \\ &= 844.84 \text{ liters s}^{-1} \end{aligned}$$

The error arising in calculating the net pumping speed $S_n$ using the molecular conductance of the pipe $C_{mT}$ (which includes the end effect) instead of using the molecular conductance of the pipe itself $C_{mp}$ is

$$\begin{aligned} \text{Error} &= 100(1061 \text{ liters s}^{-1} - 844.84 \text{ liters s}^{-1})/844.84 \text{ liters s}^{-1} \\ &= 25.59\% \end{aligned}$$

*Discussion*: This error can be avoided on subtracting as reciprocals the conductance of the aperture $C_{ma}$ to the pipe; thus, from Eq. 6.2c

$$\begin{aligned} C_{ma} &= 3.64(T/M)^{1/2} A \\ &= 3.64(296/28.98)^{1/2} \times 356.15 = 4143.15 \text{ liters s}^{-1} \end{aligned}$$

and

$$\begin{aligned} 1/C_{mp} &= 1/C_{mT} - 1/C_{ma} = 1/2130.53 \text{ liters s}^{-1} - 1/4143.15 \text{ liters s}^{-1} \\ &= 2.28 \times 10^{-4} \text{ liters s}^{-1} \\ C_{mp} &= 4387 \text{ liters s}^{-1} \end{aligned}$$

**Exercise 6.17**

What is the molecular conductance for air at $T = 23°\text{C}$ of the baffle valve shown schematically in Fig. 6B. The blocking plate $B$ can be moved to either end of the baffle tubing and sealed over the opening.

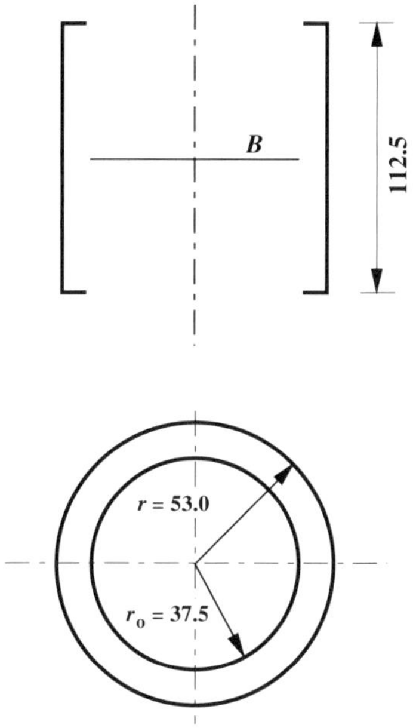


Figure 6B

**Solution 6.17**

Use Eq. 6.16a:

$$C_{mT} = 3.64(T/M)^{1/2} APr$$

Substitute

$T = 273 + 23 = 296$ K
$M = 28.98$ g mol$^{-1}$ (see Table 1.2)
$r = 5.3$ cm
$r_0 = 3.75$ cm
$A = \pi r^2 = 3.14 \times 3.75^2$ cm$^2$ = 44.156 cm$^2$
$L = 11.25$ cm

Determine $Pr$ using the graph in Fig. 6.10; then calculate

$$L/r_0 = 11.25 \text{ cm}/3.75 \text{ cm} = 3$$

$$(r/r_0)^2 = (5.3 \text{ cm}/3.75 \text{ cm})^2 = 2$$

Raise a vertical line through $L/r_0 = 3$ until it intersects the curve $(r/r_0)^2 = 2$.

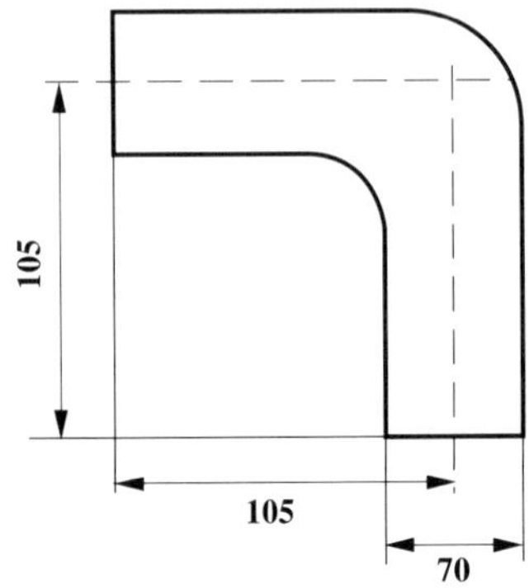


**Figure 6C**

The value of $Pr$ corresponding to the intersection point is $Pr = 0.37$. Thus

$$C_{mT} = 3.64(296/28.98)^{1/2} \times 44.156 \times 0.37 = 190 \text{ liters s}^{-1}$$

**Exercise 6.18**

Calculate the molecular conductance for air at $T = 23°C$ of the elbow shown in Fig. 6C, according to (a) the transmission probability $Pr$ through such a geometry and (b) the rule of thumb that considers the elbow as a pipe of diameter $d$ and effective length $L_{eff} < (L_{axial} + 1.33d)$.

**Solution 6.18**

(a) Use Eq. 6.16a written as

$$C_{m\,elb.} = 3.64(T/M)^{1/2} APr$$

and substitute

$T = 273 + 23 = 296$ K
$M = 28.98$ g mol$^{-1}$ (see Table 1.2)
$d = 7$ cm, $r = 3.5$ cm, $A = \pi r^2 = 3.14 \times 3.5^2$ cm$^2$ $= 38.465$ cm$^2$
$L = 10.5$ cm $+ 10.5$ cm $= 21$ cm, total length of the elbow $L/r = 21$ cm$/3.5$ cm $= 6$

To determine $Pr$, refer to Fig. 6.7. The vertical line raised through $L/r = 6$

intersects the curve of the molecular transmission probability at $Pr = 0.27$. Hence

$$C_{m\ elb.} = 364(296/28.98)^{1/2} \times 38.465 \times 0.27 = 120.82 \text{ liters s}^{-1}$$

(b) Use Eq. 6.9a written as

$$C_{m\ elb.} = 12.1 d^3 L_{eff}$$

where

$$L_{eff} < (L_{axial} + 1.33d)$$

Substitute

$d = 7$ cm
$L_{axial} = 21$ cm

Thus $L_{eff} < (21 \text{ cm} + 1.33 \times 7 \text{ cm})$ and $L_{eff} < 30.31$ cm. Taking $L_{eff} = 30$ cm, there results $C_{m\ elb} = 12.1 \times 7^3/30 = 138.34$ liters s$^{-1}$. The error in evaluating the molecular conductance of an elbow by using the rule of thumb instead of calculating it analytically is

$$\begin{aligned}\text{Error} &= 100(138.4 \text{ liters s}^{-1} - 120.82 \text{ liters s}^{-1})/120.82 \text{ liters s}^{-1} \\ &= 14.5\%\end{aligned}$$

**Exercise 6.19**

Calculate the total molecular transmission probability of the vacuum components in Fig. 6D. The elbow and the frustum are matched with no discontinuity at the interface.

**Solution 6.19**

(i) Determination of the molecular transmission probability of the elbow from plane 3 to plane 2. Consider the following dimensions of the elbow:

$a = 20.0$ cm
$b = 20.0$ cm
$r = 8.0$ cm

To determine the molecular transmission probability, calculate

$$(a + b)/r = (20.0 \text{ cm} + 20.0 \text{ cm})/8.0 \text{ cm} = 5$$

On the graph in Fig. 6.7 the vertical interrupted line raised through $(a + b)/r = 5$ intersects the molecular transmission probability curve at $Pr = 0.315$. Thus

$$Pr_{elb.\ 3-2} = 0.315$$

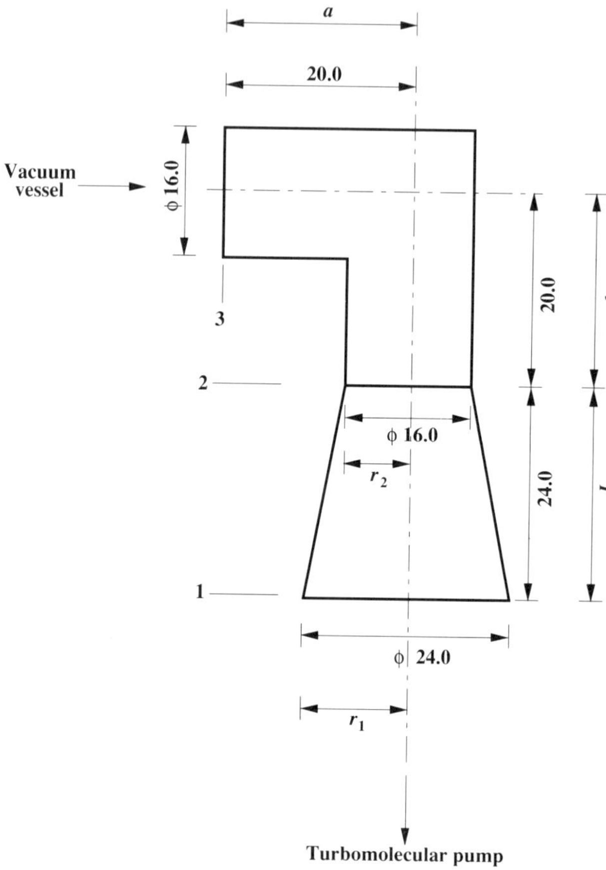


Figure 6D

(ii) Determination of the molecular transmission probability of the frustum from plane 2 to plane 1. The dimensions of the frustum are

$r_1 = 12.0$ cm
$r_2 = 8.0$ cm
$L = 24.0$ cm

For graphical determination of the molecular transmission probability of the frustum from plane 2 to plane 1, calculate

$$L/r = 24.0 \text{ cm}/8.0 \text{ cm} = 3$$

$$r_1/r_2 = 12.0 \text{ cm}/8.0 \text{ cm} = 1.5$$

On the graph in Fig. 6.18, the interrupted line starting from the intersection of curves $L/r_2 = 3$ and $r_1/r_2 = 1.5$, drawn parallel to the abscissa of the graph, intersects the vertical axis at $Pr_{1-2} = 0.29$. Thus $Pr_{\text{frust. }1-2} = 0.29$. To obtain $Pr_{\text{frust. }2-1}$, the reverse molecular transmission probability, use Eq. 6.20 written as

$$A_1 Pr_{\text{frust. }1-2} = A_2 Pr_{\text{frust. }2-1}$$

Thus

$$Pr_{\text{frust. }2-1} = (A_1/A_2) Pr_{\text{frust. }1-2}$$

Substitute

$$A_1 = \pi r_1^2 = 3.14 \times 12^2 \text{ cm}^2 = 462.16 \text{ cm}^2$$

$$A_2 = \pi r_2^2 = 3.14 \times 8^2 \text{ cm}^2 = 200.96 \text{ cm}^2$$

Hence

$$Pr_{\text{frust. }2-1} = \left(452.16 \text{ cm}^2/200.96 \text{ cm}^2\right) \times 0.29 = 0.6525$$

(iii) Determination of the total molecular transmission probability of the assembly elbow–frustum. Use Eq. 6.30 written as

$$Pr_{\text{assembly }3-1} = Pr_{\text{elb. }3-2} Pr_{\text{frust. }2-1} / (Pr_{\text{elb. }2-3} + Pr_{\text{frust. }2-1} - Pr_{\text{elb. }3-2} Pr_{\text{frust. }2-1})$$

Substitute

$$Pr_{\text{elb. }3-2} = Pr_{\text{elb. }2-3} = 0.315$$

$$Pr_{\text{frust. }2-1} = 0.6525$$

Thus

$$Pr_{\text{assembly}} = 0.315 \times 0.6525/(0.315 + 0.6525 - 0.315 \times 0.6525)$$

$$= 0.2697$$

## References

Barret, A., and Bosanquet, C. H. (1944), *Resistance of Ducts to Molecular Flow*, Imperial Chemical Industries, Ltd. (Birmingham Division), Report BR-266, November.

Bureau, A. J., Lasslet, J., and Keller, J. M. (1952), *Rev. Sci. Instrum.*, *12*, 683.

Carlsons, R. W. (1979), in *Methods of Experimental Physics*, Vol. 14, *Vacuum Physics and Technology*, R. Carlson and G. L. Weissler (eds.), Academic Press, New York.

Clausing, P. (1932), *Ann. Physik.*, *12*, 961, reprinted in *J. Vac. Sci. Technol.* (1971), *8*, 636.

Cole, R. J. (1976), *Prog. Astronaut. Aeronaut.*, *51*, 261.

Davis, D. H. (1960), *J. Appl. Phys.*, *31*, 7, 1169.

Davis, D. H. (1961), *UCRL-5952*, Lawrence Radiation Laboratory, University of California, Livermore, Ca.

Dushman, S. (1966), *Scientific Foundations of Vacuum Technique*, 2nd ed., J. M. Lafferty (ed.), Wiley, New York.

Guthrie, A. and Wakelring, R. K. (1949), *Vacuum Equipment and Techniques*, McGraw-Hill, New York.

Knudsen, M. (1909), *Ann. Physik*, *28*, 75, 999.

Knudsen, M. (1911), *Ann. Physik*, *35*, 389.

Levenson, L. L., Milleron, N., and Davis, H. D. (1960), *Transactions 7th National Symposium on Vacuum Technology*, Pergamon Press, Oxford (1961).

Levenson, L. L., Milleron, N., and Davis, H. D. (1963), *Le Vide*, *18*, 42.

Loeb, L. B. (1934), *Kinetic Theory of Gases*, McGraw-Hill, New York.

Oatley, C. W. (1957), *Br. J. Appl. Phys.*, *8*, 1, 15.

Onusic, H. (1980), *J. Vac. Sci. Technol.* *17*, *2*, 661.

Roth, A. (1990), *Vacuum Technology*, 3rd updated and expanded edition, North-Holland, Amsterdam.

Santeler, D. J. (1986), *J. Vac. Sci. Technol.* A4, *3*, 338.

Santeler, D. J., and Boeckman, M. D. (1987), *J. Vac. Sci. Technol.* A5, *4*, 2403.

Santeler, D. J., Holkeboer, H. D., Jones, W. D., and Pagano, F. (1966), *Vacuum Technology and Space Simulation*, National Aeronautics Space Administration, NASA SP-105, Washington, D.C.

Steckelmacher, W. (1978), *Vacuum*, *28*, 269.

# 7

# Steady Flow of Gas in the Transition Range

## 7.1 *Conductance of Long Cylindrical Pipes in the Transition Range*

The flow of gas in the transition range (denoted by subscript tr) is known as *Knudsen's flow*. This occurs when the mean path of molecules is of the same order of magnitude as the transverse dimensions of the pipe through which gas flows.

Knudsen (1911) expressed the conductance of a long cylindrical pipe as

$$C_{tr} = C_l + JC_{mp} \qquad (\text{e.g., cm}^3\ \text{s}^{-1}) \tag{7.1}$$

with

$$J = \left[1 + (2r/\eta)(M/RT)^{1/2}\bar{p}\right] \Big/ \left[1 + 2.47(r/\eta)(M/RT)^{1/2}\bar{p}\right] \qquad (\text{dimensionless}) \tag{7.2}$$

$$= \left[1 + 2.507\left(r/\lambda_{\bar{p}}\right)\right] \Big/ \left[1 + 3.095\left(r/\lambda_{\bar{p}}\right)\right] \qquad (\text{dimensionless}) \tag{7.2a}$$

where $C_l$ (e.g., $\text{cm}^3\ \text{s}^{-1}$) is the laminar conductance of a long straight pipe of circular cross section (see Eq. 5.12), $C_{mp}$ (e.g., $\text{cm}^3\ \text{s}^{-1}$) is the molecular conductance of a straight pipe of circular cross section (see Eq. 6.8), $r$ (e.g., cm) is the inner radius of the pipe, and $\lambda_{\bar{p}}$ (e.g., cm) is the mean free path of molecules corresponding to the average pressure $\bar{p}$ in Eq. 7.2.

Equation 7.1 can also be written (Dushman, 1966, p. 106) as

$$C_{tr} = C_{mp}\left[\left(C_l/C_{mp}\right) + J\right] \qquad (\text{e.g., cm}^3\ \text{s}^{-1}) \tag{7.3}$$

$$= C_{mp}\left(0.1472\left(r/\lambda_{\bar{p}}\right) + J\right) \tag{7.3a}$$

Thus, the transition conductance of a pipe is a function of both $J$ and $(r/\lambda_{\bar{p}})$.

Table 7.1 lists the ratio of the transition conductance to the molecular conductance of cylindrical pipes ($C_{tr}/C_{mp}$) as a function of the ratio of the

**Table 7.1**
Ratio of Transition Conductance $C_{tr}$ to Molecular Conductance $C_{mp}$ of Cylindrical Pipes for a Range of Values $r/\lambda_{\bar{p}}$[a]

| $r/\lambda_{\bar{p}}$ | $0.1472(r/\lambda_{\bar{p}})$ | $J$ | $C_{tr}/C_{mp}$ |
|---|---|---|---|
| $10^4$ | 1472.0 | 0.810 | 1472.81 |
| $10^3$ | 147.2 | 0.810 | 148.01 |
| $10^2$ | 14.72 | 0.810 | 15.53 |
| 10 | 1.472 | 0.816 | 2.288 |
| 5 | 0.736 | 0.822 | 1.558 |
| 1 | 0.147 | 0.857 | 1.004 |
| 0.5 | 0.074 | 0.885 | 0.959 |
| 0.323 | 0.047 | 0.905 | 0.952 min |
| 0.2 | 0.029 | 0.933 | 0.962 |
| 0.1 | 0.015 | 0.955 | 0.970 |
| 0.05 | 0.007 | 0.974 | 0.981 |
| 0.01 | 0.002 | 0.995 | 0.997 |
| 0 | 0 | 1.000 | 1.000 |

[a]After Dushman (1966), p. 107.

radius of the pipe $r$ to the mean free path $\lambda_{\bar{p}}$ corresponding to the average pressure $\bar{p}$ in the pipe.

On replacing the values of $C_l$ (Eq. 5.12), $C_{mp}$ (Eq. 6.9), and $J$ (Eq. 7.2) in Eq. 7.1, the conductance to transition flow of long pipes (Van Atta, 1965, pp. 37–39) becomes

$$C_{tr}(L/d^3) = 3.269 \times 10^{-2}(\bar{p}d/\eta) + 3.81(T/M)^{1/2} \times\left[1 + 0.147(M/T)^{1/2}(\bar{p}d/\eta)\right] \Big/ \left[1 + 0.181(M/T)^{1/2}(\bar{p}d/\eta)\right] \quad \text{(liters s}^{-1}\text{)} \quad (7.4)$$

with $L$ (cm), $d$ (cm), $\bar{p}$ (Torr), $\eta$ (poise), $T$ (K), and $M$ (g mol$^{-1}$). For air at 23°C (Van Atta, 1965, p. 37)

$$C_{tr} = 178.7\bar{p}(d^4/L) + 12.12[(1 + 252.1\bar{p}d)/(1 + 311.7\bar{p}d)](d^3/L) \quad \text{(liters s}^{-1}\text{)} \quad (7.4a)$$

The minimum conductance in the transition flow occurs when

$$\bar{p}_{min}d = 5.47(T/M)^{1/2}\eta \quad \text{(Torr cm)} \quad (7.5)$$

For air at $T = 23°C$

$$\bar{p}_{min} d = 3.10 \times 10^{-3} \qquad \text{(Torr cm)} \tag{7.5a}$$

and

$$\lambda_{min}/d = 1.59 \qquad \text{(dimensionless)} \tag{7.5b}$$

## 7.2 Transition Pressure in Long Cylindrical Pipes

$$\bar{p}_{tr} d = 95.7(T/M)^{1/2}\eta \qquad \text{(Torr cm)} \tag{7.6}$$

For air at $T = 23°C$:

$$\bar{p}_{tr} d = 5.43 \times 10^{-2} \qquad \text{(Torr cm)} \tag{7.6a}$$

## 7.3 Maximum and Minimum Pressures of the Transition Range

$$\bar{p}_{tr\,max} d = 942(T/M)^{1/2}\eta \qquad \text{(Torr cm)} \tag{7.7}$$

For air at $T = 23°C$:

$$\bar{p}_{tr\,max} d = 5.35 \times 10^{-1} \qquad \text{(Torr cm)} \tag{7.7a}$$

$$\bar{p}_{tr\,min} d = 10(T/M)^{1/2}\eta \qquad \text{(Torr cm)} \tag{7.8}$$

For air at $T = 23°C$:

$$\bar{p}_{tr\,min} d = 5.7 \times 10^{-3} \qquad \text{(Torr cm)} \tag{7.8a}$$

## 7.4 Maximum and Minimum Mean Free Path in the Transition Range

$$\lambda_{tr\,max} = 9.09 \times 10^{-3} d \qquad \text{(cm)} \tag{7.9}$$

$$\lambda_{tr\,min} = 9.09 \times 10^{-1} d \qquad \text{(cm)} \tag{7.10}$$

For many practical vacuum systems, it is sufficient to limit calculations to viscous or molecular flow, since the transition flow occurs for a very short period of time between these flow regimes.

Detailed equations for flow in the transition range have been published by Thomson and Owens (1975). For the transition flow through an aperture, see De Muth and Watson (1986).

## Exercises

**Exercise 7.1**

Calculate the conductance for air at $T = 23°C$ of a cylindrical pipe 3 m long, 2 cm in inner diameter, in the transition range, assuming that the average pressure is 150 Torr. What is the value of the minimum conductance of the pipe in this range?

**Solution 7.1**

Use Eq. 7.4a:

$$C_{tr} = 178.7\bar{p}(d^4/L) + 12.12[(1 + 252.1\bar{p}d)/(1 + 311.7\bar{p}d)](d^3/L)$$

Substitute

$\bar{p} = 150$ Torr
$d = 2$ cm
$L = 300$ cm

Thus

$$\begin{aligned} C_{tr} &= 178.7 \times 150(2^4/300) \\ &\quad + 12.12[(1 + 252.1 \times 150 \times 2)/ \\ &\quad (1 + 311.7 \times 150 \times 2)](2^3/300) \\ &= 373 \qquad (\text{liters s}^{-1}) \end{aligned}$$

The minimum conductance of the pipe occurs when (Eq. 7.5a)

$$\bar{p}_{min}d = 3.10 \times 10^{-3} \text{ Torr cm}$$

With $d = 2$ cm

$$\bar{p}_{min} = 3.10 \times 10^{-3}/2 = 1.52 \times 10^{-3} \text{ Torr}$$

At that pressure (Eq. 7.5b)

$$\lambda_{min}/d = 1.59$$

With $d = 2$ cm

$$\lambda_{min} = 1.59/2 = 0.795 \text{ cm}$$

**Exercise 7.2**

What is the transition pressure for air in the pipe specified in Exercise 7.1?

**Solution 7.2**

Use Eq. 7.6a:

$$\bar{p}_{tr}d = 5.43 \times 10^{-2} \text{ Torr cm}$$

Substitute $d = 2$ cm. Thus

$$\bar{p}_{tr} = 5.43 \times 10^{-2}/2 = 2.715 \text{ Torr}$$

## References

De Muth, S. F., and Watson, J. S. (1986), *J. Vac. Sci. Technol.* *A4*, *3*, 344.

Dushman, S. (1966), in *Scientific Foundations of Vacuum Technique*, 2nd ed., J. M. Lafferty (ed.), Wiley, New York.

Knudsen, M. (1911), *Ann. Physik.* *35*, 389.

Van Atta, C. M. (1965), *Vacuum Science and Engineering*, McGraw-Hill, New York.

Thomson, S. L., and Owens, W. R. (1975), *Vacuum* *25*, 151.

# 8

# Gas Load

## 8.1 *Sources of Gas in Vacuum Systems*

In addition to the residual gas from the initial atmosphere in a pumped-down vessel the following sources of gas may be present in a vacuum system:

The gas contributed by the outgassing of materials
The gas generated by the evaporation of the materials processed or the measuring instruments (e.g. gauges)
The gas resulting from leakage
The gas that penetrates through permeation

These sources of gas are schematically illustrated in Fig. 8.1.

## 8.2 *Evaluation of the Gas Load*

Depending on the degree of rarefaction reached in a vacuum vessel, the surface finish and previous history of exposure to a variety of atmospheric conditions of the vessel, cleaning methods, temperature and nature of the materials processed, and other factors, the gas load $Q_G$ is expressed as

$$Q_G = Q_D + Q_E + Q_L + Q_{Per} \qquad (\text{e.g., Pa m}^3\ \text{s}^{-1}) \qquad (8.1)$$

where subscripts to the gas-flow rates $Q$ refer to outgassing (D), evaporation (E), leaks (L), and permeation (Per).

Since *the gas load limits the ultimate pressure of the vacuum system*, the evaluation of the gas load is one of the basic considerations in calculation of the time required to reach a certain pressure.

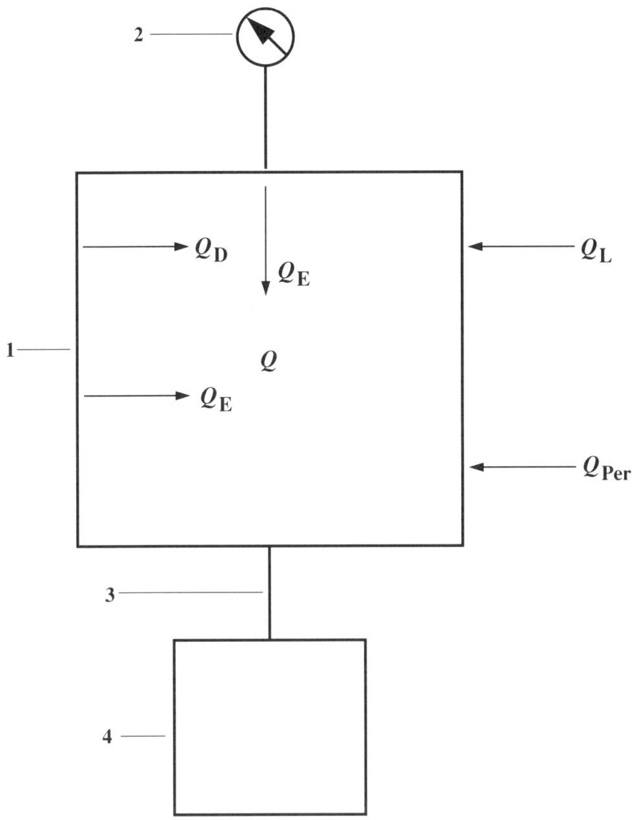


**Figure 8.1** Sources of gas in a pumped system: 1—vacuum vessel, 2—vacuum gauge, 3—pipe, 4—vacuum pump, $Q$—residual gas from the initial atmosphere, $Q_D$—gas contributed by the outgassing of materials, $Q_E$—gas produced by the evaporation of materials, $Q_L$—gas resulting from leakage, $Q_{Per}$—gas that permeates the vacuum vessel.

### 8.2.1 Outgassing (Surface)

$$Q_D = q_D A \qquad \text{(e.g., mbar s}^{-1}\text{)} \tag{8.2}$$

where $q_D$ (e.g., mbar liters $s^{-1}$ $cm^{-2}$) is the specific outgassing rate of the material considered, and $A$ ($cm^2$) is the total outgassing area.

The specific outgassing rate $q_D$ is dependent on temperature, surface treatment, and previous history. Tables 8.1–8.4 list the specific outgassing rates of unbacked metal, elastomers, polymers, and glass and ceramics.

**Table 8.1**
Specific Outgassing Rates of Unbaked Metals after 1 h and 10 h, Respectively, from Startup of Pumpdown

| Metal | $q_{Dm}$ (1 h) (Torr liters $s^{-1}$ $cm^{-2}$) | $\theta^a$ 1 h | $q_{Dm}$ (10 h) (Torr liters $s^{-1}$ $cm^{-2}$) | $\theta^a$ 10 h | Ref. |
|---|---|---|---|---|---|
| Mild steel | $5.4 \times 10^{-7}$ | 1 | $5.0 \times 10^{-8}$ | 1 | Dayton (1959) |
| Slightly rusty | $6.0 \times 10^{-7}$ | 1 | $1.3 \times 10^{-8}$ | 1 | |
| Chromium-plated, polished | $1.0 \times 10^{-8}$ | | $9.08 \times 10^{-9}$ | | |
| Aluminum-spray-coated | $6.0 \times 10^{-8}$ | 0.75 | $1.0 \times 10^{-8}$ | 0.75 | |
| Stainless steel | $9.0 \times 10^{-8}$ | 0.7 | $2.0 \times 10^{8}$ | 0.75 | |
| Steel | | | | | Schram (1963) |
| Descaled | $3.07 \times 10^{-7}$ | 0.6 | $2.95 \times 10^{-7}$ | 0.7 | |
| Chromium-plated, fresh | $7.05 \times 10^{-9}$ | 1 | $5.8 \times 10^{-10}$ | 1 | |
| Chromium-plated, polished | $9.1 \times 10^{-9}$ | 1 | $8.0 \times 10^{-10}$ | 1 | |
| Nickel-plated, fresh | $4.24 \times 10^{-9}$ | 0.9 | $4.94 \times 10^{-10}$ | 0.9 | |
| Nickel-plated | $2.76 \times 10^{-9}$ | 1.1 | $2.23 \times 10^{-10}$ | 1.1 | |
| Chemically nickel-plated, fresh | $8.3 \times 10^{-9}$ | 1 | $7.05 \times 10^{-10}$ | 1 | |
| Chemically nickel-plated, polished | $5.22 \times 10^{-9}$ | 1 | $4.6 \times 10^{-10}$ | 1 | |
| Stainless steel | | | | | |
| ICN 472, fresh | $1.35 \times 10^{-8}$ | 0.9 | $1.47 \times 10^{-9}$ | 0.9 | |
| ICN sanded | $8.28 \times 10^{-9}$ | 1.2 | $1.04 \times 10^{-9}$ | 0.8 | |
| NS 226, mechanically polished | $1.44 \times 10^{-8}$ | 1.3 | $1.35 \times 10^{-9}$ | 1.9 | |
| Stainless steel | | | | | Schram (1963) |
| NS 226 electro polished | $4.28 \times 10^{-5}$ | 1 | $4.28 \times 10^{-10}$ | 1 | |
| Aluminum | | | | | |
| Fresh | $6.3 \times 10^{-9}$ | 1 | $6.0 \times 10^{-10}$ | 1 | |
| Anodized 2-$\mu$m pores | $2.76 \times 10^{-7}$ | 0.9 | $3.22 \times 10^{-8}$ | 0.9 | |
| 3 h in air | $6.65 \times 10^{-9}$ | 1.9 | $4.75 \times 10^{-10}$ | 0.9 | |
| Degreased 24 h in air | $4.14 \times 10^{-9}$ | 0.9 | $3.06 \times 10^{-10}$ | 0.9 | |
| Copper | | | | | |
| Fresh | $4.0 \times 10^{-8}$ | 1 | $4.15 \times 10^{-9}$ | 1 | |
| Mechanically polished | $3.5 \times 10^{-9}$ | 1 | $3.56 \times 10^{-10}$ | 1 | |
| OFHC[b] fresh | $1.88 \times 10^{-8}$ | 1.3 | $1.26 \times 10^{-9}$ | 1.3 | |
| OFHC[b] mechanically polished | $1.9 \times 10^{-9}$ | 1.1 | $1.63 \times 10^{-10}$ | 1.1 | |
| Gold wire, fresh | $1.58 \times 10^{-7}$ | 2.1 | $5.1 \times 10^{-10}$ | 1 | |
| Molybdenum | $5.22 \times 10^{-9}$ | 1 | $3.67 \times 10^{-10}$ | 1 | |
| Zinc | $2.21 \times 10^{-7}$ | 1.4 | $3.22 \times 10^{-8}$ | 0.8 | |
| Titanium | $4.0 \times 10^{-9}$ | 1 | $3.68 \times 10^{-10}$ | 1 | |

[a] $\theta$ is the slope on a log–log plot of the outgassing rate versus time. The negative sign is omitted.
[b] Oxygen-free high conductivity.

**Table 8.2**
Specific Outgassing Rates of Elastomers

| Material | $q_{De}$(1 h) (Torr liters $s^{-1}$ $cm^{-2}$) | $\theta$[a] 1 h | $q_{De}$(10 h) (Torr liters $s^{-1}$ $cm^{-2}$) | $\theta$[a] 10 h | Ref. |
|---|---|---|---|---|---|
| Butyl DR 41 | $1.5 \times 10^{-8}$ | 0.68 | $4.0 \times 10^{-9}$ | 0.64 | Blears et al. (1960) |
| Natural crepe | $7.3 \times 10^{-8}$ | 0.7 | $4.0 \times 10^{-9}$ | 0.65 | |
| Neoprene | $3.0 \times 10^{-7}$ | 0.4 | $1.8 \times 10^{-7}$ | 0.4 | |
| Nygon | $1.3 \times 10^{-7}$ | 0.5 | $6.5 \times 10^{-8}$ | 0.6 | |
| Perbunan | $3.5 \times 10^{-8}$ | 0.3 | $2.2 \times 10^{-8}$ | 0.5 | |
| Viton A, fresh | $1.14 \times 10^{-8}$ | 0.8 | — | — | |

[a] $\theta$ is the slope on a log–log plot of the outgassing rate versus time. The negative sign is omitted.

**Table 8.3**
Specific Outgassing Rates of Polymers

| Material | $q_{Dpoly.}$ (1 h) (Torr liters $s^{-1}$ $cm^{-2}$) | $\theta$[a] 1 h | $q_{Dpoly.}$(10 h) (Torr liters $s^{-1}$ $cm^{-2}$) | $\theta$[a] 10 h | Ref. |
|---|---|---|---|---|---|
| Araldite D | $1.9 \times 10^{-8}$ | 0.3 | $1.225 \times 10^{-8}$ | 0.5 | Geller (1958) |
| Araldite F | $1.5 \times 10^{-8}$ | 0.5 | $7.3 \times 10^{-9}$ | 0.5 | |
| Plexiglass | $3.1 \times 10^{-8}$ | 0.4 | $1.8 \times 10^{-8}$ | 0.4 | |
| Metyl-methacrylate | $4.2 \times 10^{-8}$ | 0.9 | $1.4 \times 10^{-8}$ | 0.57 | Blears et al. (1960) |
| PMMA (poly-methylmethacylate | $5.3 \times 10^{-7}$ to $1.5 \times 10^{-6}$ | 1.6 | $1.8 \times 10^{-8}$ to $3.9 \times 10^{-8}$ | 1.7 | Rosenblum (1986) |
| PTFE (poly-fluoreth-ylene | $3.0 \times 10^{-9}$ | 0.45 | $1.5 \times 10^{-9}$ | 0.56 | Dayton (1959) |
| Teflon | $6.5 \times 10^{-8}$ | 0.5 | $2.5 \times 10^{-8}$ | 0.2 | Power and Crawly (1960) |

[a] $\theta$ is the slope on a log–log plot of the outgassing rate versus time. The negative sign is omitted.

**Table 8.4**
Specific Outgassing Rates of Glasses and Ceramics

| Material | $q_{Dgs}$ (1 h) (Torr liters $s^{-1}$ $cm^{-2}$) | $\theta$[a] 1 h | $q_{Dgs}$ (10 h) (Torr liters $s^{-1}$ $cm^{-2}$) | $\theta$[a] 10 h | Ref. |
|---|---|---|---|---|---|
| Pyrex, fresh | $7.35 \times 10^{-9}$ | 1.1 | $5.5 \times 10^{-10}$ | 1.7 | Schram (1963) |
| Pyrex, 1 month in air | $1.16 \times 10^{-9}$ | 0.9 | $1.6 \times 10^{-10}$ | 0.7 | |
| Steatite | $9.0 \times 10^{-8}$ | 1 | $9.5 \times 10^{-9}$ | 1 | |

[a] $\theta$ is the slope on a log–log plot of the outgassing rate versus time. The negative sign is omitted.

**Table 8.5**
Conversion Factors for Outgassing Rates in Different Systems of Units

| | $Pa\ m^3\ s^{-1}\ m^{-2}$ or $W\ m^{-2}$ | mbar liters $s^{-1}\ cm^{-2}$ | Torr liters $s^{-1}\ cm^{-2}$ | $\mu l\ s^{-1}\ cm^{-2}$ | $\mu\ ft^3\ s^{-1}\ cm^{-2}$ |
|---|---|---|---|---|---|
| $Pa\ m^3\ s^{-1}\ m^{-2}$ or $W\ m^{-2}$ | 1.0 | $1.0 \times 10^{-3}$ | $7.5 \times 10^{-4}$ | $7.5 \times 10^{-1}$ | $2.65 \times 10^{-2}$ |
| mbar liters $s^{-1}\ cm^{-2}$ | $1.0 \times 10^{-3}$ | 1.0 | $7.5 \times 10^{-1}$ | $7.5 \times 10^{2}$ | $2.65 \times 10$ |
| Torr liters $s^{-1}\ cm^{-2}$ | $1.33 \times 10^{3}$ | 1.33 | 1.0 | $1.0 \times 10^{3}$ | $3.53 \times 10$ |
| $\mu l\ s^{-1}\ cm^{-2}$ | 1.33 | $1.33 \times 10^{-3}$ | $1.0 \times 10^{-3}$ | 1.0 | $3.53 \times 10^{-2}$ |
| $\mu\ ft^3\ s^{-1}\ cm^{-2}$ | $3.78 \times 10$ | $3.78 \times 10^{-2}$ | $2.83 \times 10^{-2}$ | $2.83 \times 10$ | 1.0 |

The conversion factors for specific outgassing rates as well as for gas quantities, in different system of units, are listed in Tables 8.5 and 8.6.

To convert the amount of gas in pressure–volume units to mass units at STP, use the following equations:

$$1\ \text{Pa m}^3(\text{J}) = 4.406 \times 10^{-4}\ M \quad (\text{g}) \tag{8.3}$$

$$1\ \text{mbar} = 4.406 \times 10^{-5}\ M \quad (\text{g}) \tag{8.4}$$

$$1\ \text{Torr} = 5.784 \times 10^{-5}\ M \quad (\text{g}) \tag{8.5}$$

$$1\ \text{atm cm}^3 = 4.464 \times 10^{-5}\ M \quad (\text{g}) \tag{8.6}$$

To convert the amount of gas (g) at STP to another temperature ($T_K$), multiply by 273 K/$T_K$.

Dayton (1960, 1962, 1963) derived the complete theory of outgassing.

**Table 8.6**
Conversion Factors for Gas Quantities in Different Systems of Units

| | $Pa\ m^3$ or J | mbar liters | Torr liters | $\mu l$ | $\mu ft^3$ |
|---|---|---|---|---|---|
| $Pa\ m^3$ or J | 1 | 10 | 7.50 | $7.50 \times 10^{3}$ | $2.65 \times 10^{2}$ |
| mbar liters | $1.0 \times 10^{-1}$ | 1 | $7.50 \times 10^{-1}$ | $7.50 \times 10^{2}$ | $2.65 \times 10$ |
| Torr liters | $1.33 \times 10^{-1}$ | 1.33 | 1 | $1.0 \times 10^{3}$ | $3.53 \times 10$ |
| $\mu l$ | $1.33 \times 10^{-4}$ | $1.33 \times 10^{-3}$ | $1.0 \times 10^{-3}$ | 1 | $3.53 \times 10^{-2}$ |
| $\mu ft^3$ | $3.78 \times 10^{-3}$ | $3.7 \times 10^{-2}$ | $2.83 \times 10^{-2}$ | $2.83 \times 10$ | 1 |

### 8.2.2 Evaporation

Evaporation is denoted by (subscript E).

*a. Vapor pressure of substances.* The vapor pressure of metals and carbon in the range below 1 Torr can be calculated (Dushman, 1966, p. 692) as

$$\log p_E = A' - B'/T \qquad (\mu\text{m Hg}) \tag{8.7}$$

where $A'$ and $B'$ are constants and $T$ is in kelvins (K).

The values of constants $A'$ and $B'$ for metals and carbon are listed in Table 8.7.

**Table 8.7**

Constants in Eq. 8.7 for the Calculation of Vapor Pressure[a]

| Metal | $A'$ | $10^{-3}B'$ |
|---|---|---|
| Si | 12.72 | 21.30 |
| Ti | 12.50 | 23.23 |
| Zr | 12.33 | 30.26 |
| Th | 12.52 | 28.44 |
| Ge | 11.71 | 18.03 |
| Sn | 10.88 | 14.87 |
| Pb | 10.77 | 9.71 |
| V | 13.07 | 25.72 |
| Cb | 14.37 | 40.40 |
| Ta | 13.04 | 40.21 |
| $Sb_2$ | 11.15 | 8.63 |
| Bi | 11.18 | 9.53 |
| Cr | 12.94 | 20.00 |
| Mo | 11.64 | 30.85 |
| W | 12.40 | 40.68 |
| U | 11.59 | 23.31 |
| Mn | 12.14 | 13.74 |
| Fe | 12.44 | 19.97 |
| Co | 12.70 | 12.11 |
| Ni | 12.75 | 20.96 |
| Ru | 13.50 | 33.80 |
| Rh | 12.94 | 27.72 |
| Pd | 11.78 | 19.71 |
| Os | 13.59 | 37.00 |
| Ir | 13.07 | 31.23 |
| Pt | 12.53 | 27.28 |

[a]After Dushman 1966, p. 700.

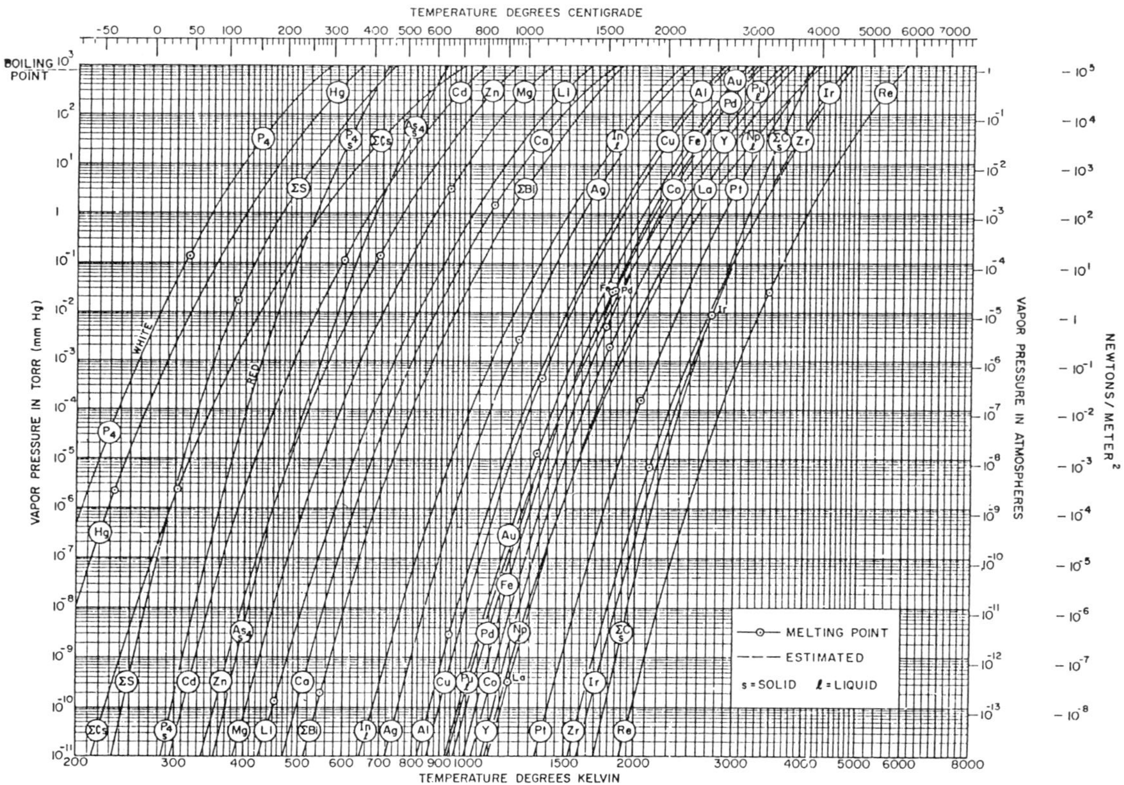


**Figure 8.2** Vapor pressure data for the solid and liquid elements. [After Honig and Kramer (1969) and courtesy of the General Electric Company.]

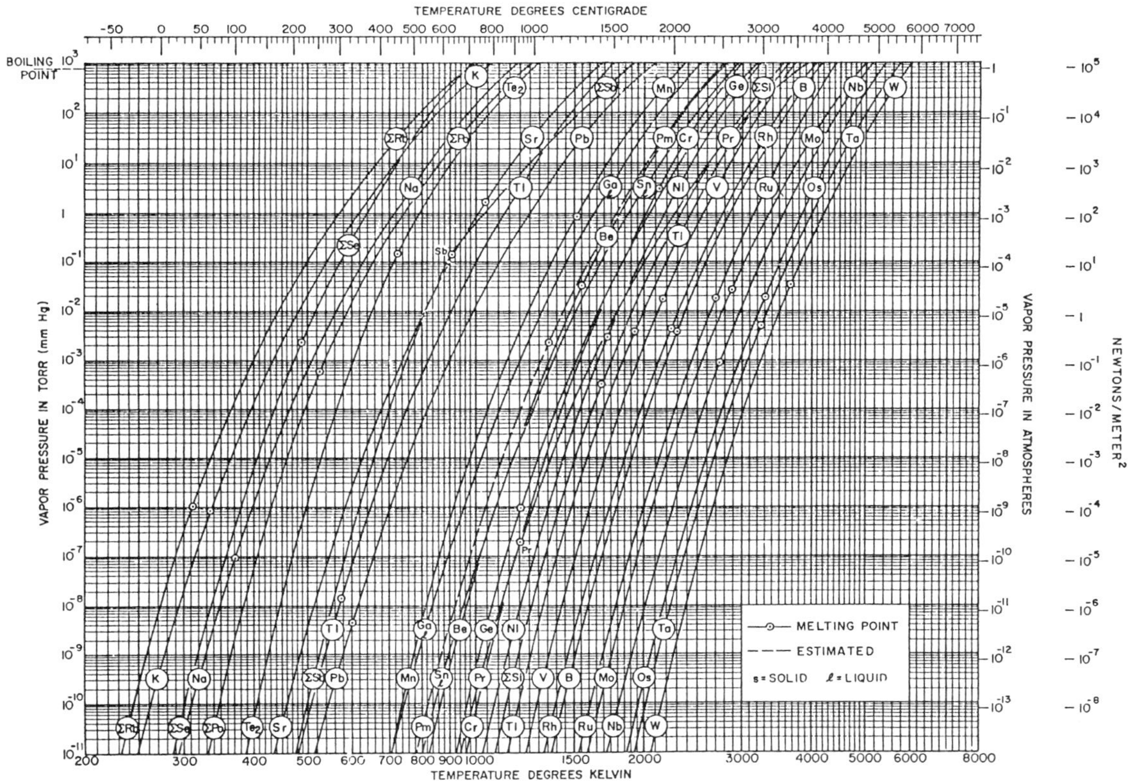


**Figure 8.3** Vapor pressure data for the solid and liquid elements. [After Honig and Kramer (1969) and courtesy of the General Electric Company.]

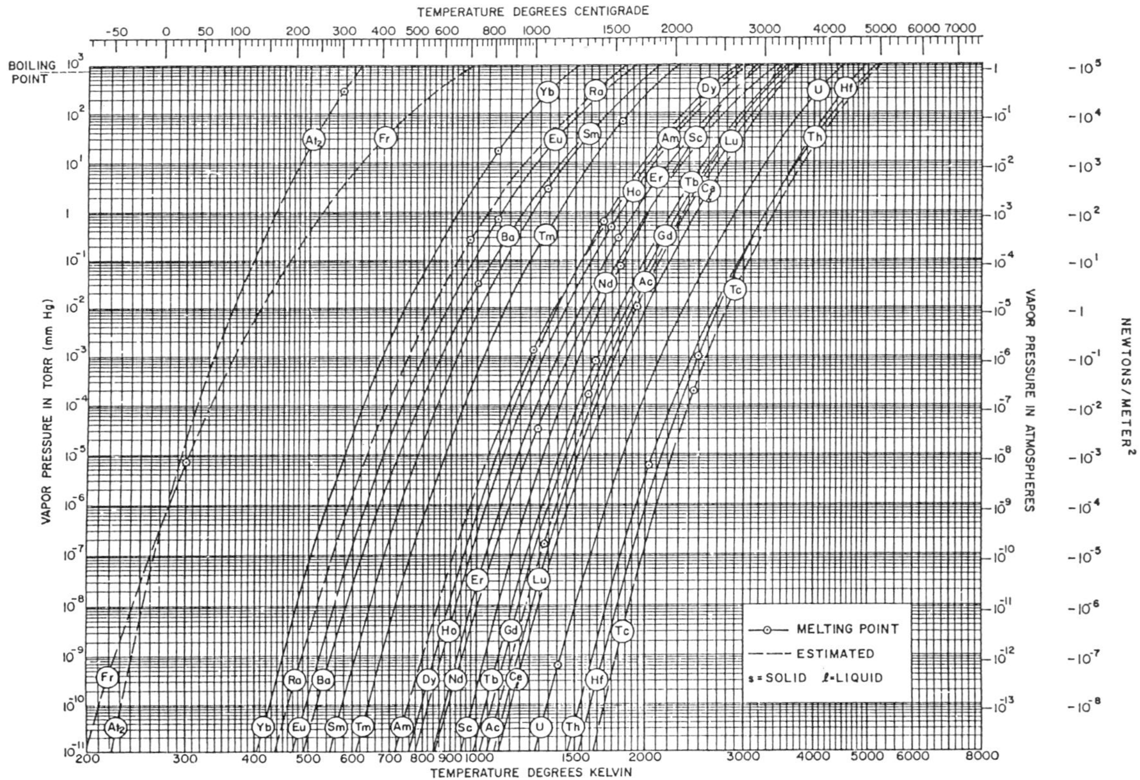


**Figure 8.4** Vapor pressure data for the solid and liquid elements. [After Honig and Kramer (1969) and courtesy of the General Electric Company.]

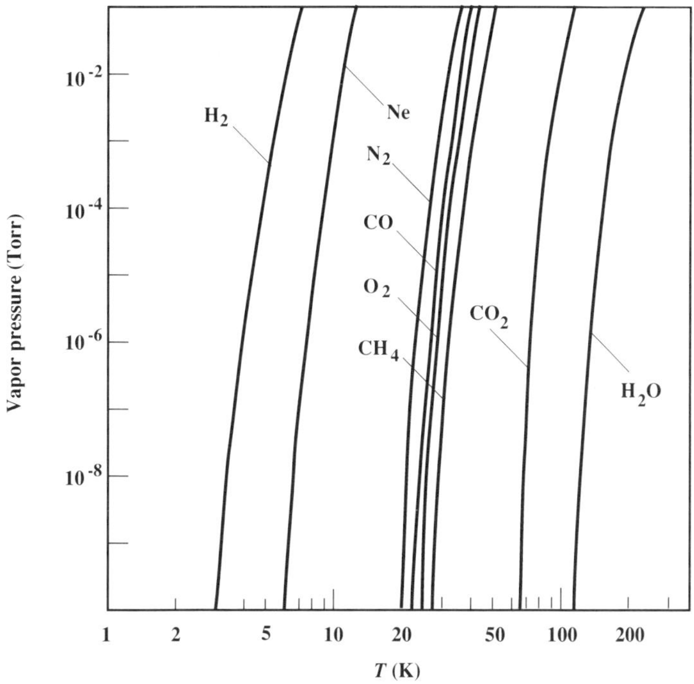


**Figure 8.5** Vapor pressure of common gases. [After Honig and Hook (1960) and courtesy of the General Electric Company.]

More data on the vapor pressure for solid and liquid elements, some common gases, and water vapor are plotted in Figs. 8.2–8.6.

The pressure of saturated water vapor in the temperature range $-15$ to 100°C is listed in Table 8.8.

*b. Evaporation rate.* At equilibrium between the solid or liquid and gaseous phases, i.e., zero net flow, the rate of evaporation $\dot{W}$ equals the rate of condensation. Thus

$$\dot{W} = Gs_f \qquad (\text{g cm}^{-2}\ \text{s}^{-1}) \tag{8.8}$$

$$= 5.834 p_E (M/T)^{1/2} s_f \tag{8.8a}$$

where G is the mass of gas incident on unit area per unit time (see Eq. 3.14e),

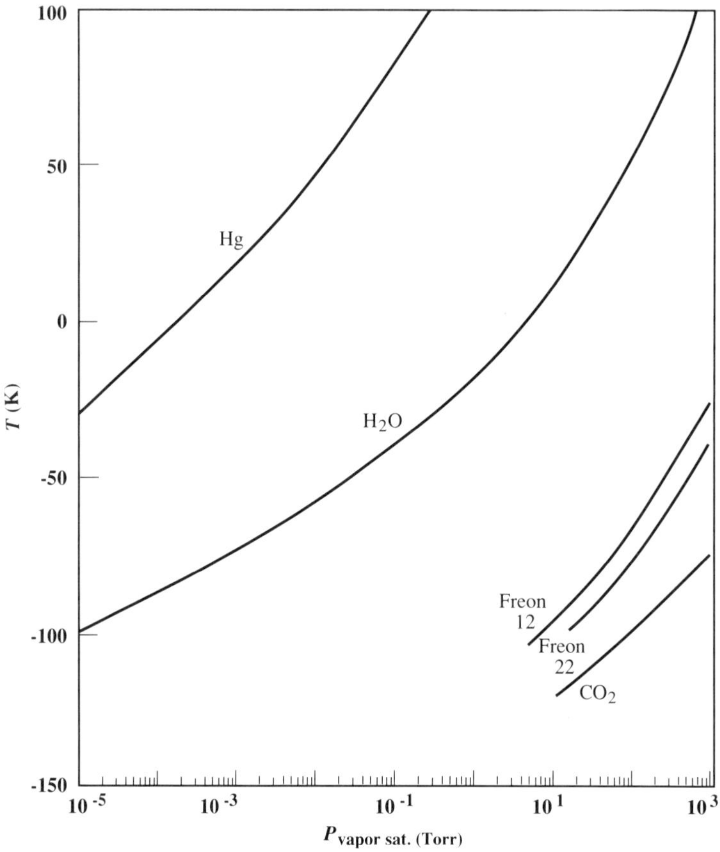


**Figure 8.6** Vapor pressure for mercury, Feon 12, Freon 22, $CO_2$ and $H_2O$. [After Dalafosse and Mongodin (1961).]

$p_E$ (Torr) is the vapor pressure of the material, and $s_f$ (dimensionless) is the sticking coefficient, with M (g mol$^{-1}$) and $T$ (K).

The rate of evaporation $\dot{W}$ (g cm$^{-2}$ s$^{-1}$) for metals and carbon, corresponding to vapor pressures $p_E$ ($\mu$m Hg) and temperatures (°C) are listed in Table 8.9.

The specific evaporation rate $q_E$ of materials, assuming the sticking coefficient $s_f = 1$, is

$$q_E = \dot{B}(p_E - p) \qquad (\text{e.g., Torr s}^{-1}\ \text{cm}^{-2}) \tag{8.9}$$

$$= 3.639(T/M)^{1/2}(p_E - p) \tag{8.9a}$$

**Table 8.8**

Pressure $p_{sv}$ of Saturated Water Vapor in the Range $T = -15$ to $\pm 100°C$

| $T$ (°C) | $p_{sv}$ (Torr)[a] | $p_{sv}$ (mbar)[b] | $T$ (°C) | $p_{sv}$ (Torr)[a] | $p_{sv}$ (mbar)[b] |
|---|---|---|---|---|---|
| −15 | 1.436 | 1.915 | 28 | 28.349 | 37.796 |
| −14 | 1.560 | 2.079 | 29 | 30.043 | 40.054 |
| −13 | 1.691 | 2.254 | 30 | 31.824 | 42.429 |
| −12 | 1.834 | 2.445 | 31 | 33.695 | 44.923 |
| −11 | 1.987 | 2.649 | 32 | 35.663 | 47.547 |
| −10 | 2.149 | 2.865 | 33 | 27.729 | 50.301 |
| −9 | 2.326 | 3.101 | 34 | 39.898 | 53.193 |
| −8 | 2.514 | 3.352 | 35 | 41.175 | 54.895 |
| −7 | 2.715 | 3.620 | 36 | 44.563 | 59.412 |
| −6 | 2.931 | 3.908 | 37 | 47.067 | 62.750 |
| −5 | 3.163 | 4.217 | 38 | 49.692 | 66.250 |
| −4 | 3.410 | 4.456 | 39 | 52.442 | 69.917 |
| −3 | 3.673 | 4.897 | 40 | 55.324 | 73.359 |
| −2 | 3.956 | 5.724 | 41 | 58.34 | 77.780 |
| −1 | 4.258 | 5.677 | 42 | 61.50 | 81.993 |
| 0 | 4.579 | 6.105 | 43 | 64.80 | 86.393 |
| 1 | 4.926 | 6.567 | 44 | 68.26 | 91.006 |
| 2 | 5.294 | 7.058 | 45 | 71.88 | 95.832 |
| 3 | 5.585 | 7.579 | 46 | 75.65 | 100.86 |
| 4 | 6.101 | 8.134 | 47 | 79.60 | 106.12 |
| 5 | 6.543 | 8.723 | 48 | 83.71 | 116.60 |
| 6 | 7.013 | 9.350 | 49 | 88.02 | 117.35 |
| 7 | 7.513 | 10.017 | 50 | 92.51 | 123.34 |
| 8 | 8.045 | 10.276 | 51 | 97.20 | 129.58 |
| 9 | 8.609 | 11.478 | 52 | 102.09 | 136.11 |
| 10 | 9.209 | 12.278 | 53 | 107.20 | 242.92 |
| 11 | 9.844 | 13.124 | 54 | 112.51 | 150.00 |
| 12 | 10.518 | 14.023 | 55 | 118.04 | 157.37 |
| 13 | 11.231 | 14.973 | 56 | 123.80 | 165.05 |
| 14 | 11.987 | 15.981 | 57 | 129.82 | 173.08 |
| 15 | 12.788 | 17.049 | 58 | 136.08 | 181.43 |
| 16 | 13.634 | 18.177 | 59 | 142.60 | 190.12 |
| 17 | 14.530 | 19.372 | 60 | 149.38 | 199.16 |
| 18 | 15.477 | 20.634 | 61 | 156.43 | 208.56 |
| 19 | 16.477 | 21.968 | 62 | 163.77 | 218.34 |
| 20 | 17.535 | 23.378 | 63 | 171.38 | 228.49 |
| 21 | 18.650 | 24.865 | 64 | 179.31 | 239.06 |
| 22 | 19.827 | 26.434 | 65 | 187.54 | 250.03 |
| 23 | 21.068 | 28.088 | 66 | 196.09 | 261.43 |
| 24 | 22.377 | 29.834 | 67 | 204.96 | 273.26 |
| 25 | 23.756 | 31.672 | 68 | 214.17 | 285.54 |
| 26 | 25.209 | 33.609 | 69 | 223.73 | 298.28 |
| 27 | 26.739 | 35.649 | 70 | 233.7 | 311.57 |

*continues*

**Table 8.8** *Continued*

| $T$ (°C) | $p_{sv}$ (Torr)[a] | $p_{sv}$ (mbar)[b] | $T$ (°C) | $p_{sv}$ (Torr)[a] | $p_{sv}$ (mbar)[b] |
|---|---|---|---|---|---|
| 71 | 243.9 | 325.17 | 86 | 450.9 | 601.15 |
| 72 | 254.6 | 339.44 | 87 | 468.7 | 624.88 |
| 73 | 265.7 | 354.24 | 88 | 487.1 | 649.41 |
| 74 | 277.2 | 369.57 | 89 | 506.1 | 674.74 |
| 75 | 289.1 | 385.44 | 90 | 525.76 | 700.96 |
| 76 | 301.4 | 401.83 | 91 | 546.05 | 728.00 |
| 77 | 314.1 | 418.77 | 92 | 566.99 | 755.92 |
| 78 | 327.3 | 436.36 | 93 | 488.60 | 784.74 |
| 79 | 341.0 | 454.63 | 94 | 610.90 | 814.47 |
| 80 | 355.1 | 473.43 | 95 | 633.90 | 845.13 |
| 81 | 369.7 | 492.89 | 96 | 657.62 | 876.75 |
| 82 | 384.9 | 513.16 | 97 | 682.07 | 909.35 |
| 83 | 400.6 | 534.09 | 98 | 707.27 | 942.95 |
| 84 | 416.8 | 555.69 | 99 | 733.24 | 977.57 |
| 85 | 433.6 | 578.09 | 100 | 760.00 | 1013.25 |

[a]Adapted from CRC (1986–1987), *Handbook of Chemistry and Physics*, pp. D189–D190.
[b]Calculated as 1 Torr = 1.333224 mbar.

where $B$ (cm$^3$ cm$^{-2}$ s$^{-1}$) is the volume of gas impinging on unit area per unit time (Eq. 3.16b), with $T$ (K), M (g mol$^{-1}$), $p_E$ (Torr), and $p$ (Torr) the pressure in the vacuum vessel that confines the evaporating material.

If the area of the material that evaporates is $A$ (cm$^2$), the flow rate $Q_E$ due to evaporation is

$$Q_E = 3.639(T/M)^{1/2}(p_E - p)A \qquad \text{(Torr liters s}^{-1}\text{)} \qquad (8.10)$$

### 8.2.3 Leakage

Gas can leak in a pumped-down vessel from external (atmosphere) or internal (gas trapped within limited volumes) sources (see Fig. 8.7). *Leaks are steadystate gas loads, which limit the ultimate pressure of a system.*

*a. Leakage (subscript L) from sources outside the vacuum system (true leaks subscript Lt).*

Refer to Fig. 8.7. The ultimate pressure $p_{us}$ that can be reached in a vacuum system in the presence of an external leak (see Fig. 8.7) is

$$p_{us} = Q_{Lt}/S_n \qquad \text{(e.g., mbar)} \qquad (8.11)$$

$$= (Q_{Lt}/V_1)\tau \qquad (8.11a)$$

where $Q_{Lt}$ (e.g., mbar liters $s^{-1}$) is the flow rate of gas at a specified temperature through the leak, $S_n$ (e.g., liters $s^{-1}$) is the net pumping speed at the port of the vacuum vessel, $V_1$ (liters) is the volume of the vessel, and $\tau$ (e.g., s) represents the time constant of the vacuum vessel (see Section 10.1.4).

If the leak of gas from atmospheric pressure occurs through a seal, the flow is expressed in pressure–volume units per second, multiplied by the sealing width and divided by the seal length, e.g., mbar liters $s^{-1}$ × mm (sealing width)/mm (sealing length).

The usual values for the flow rates due to true leaks (Santeler et al., 1966, p. 75) are

| | | |
|---|---|---|
| $Q_{Lt}$ approx. $> 10^{-5}$ | (Torr liters $s^{-1}$) | laminar flow |
| $Q_{Lt}$ approx. $< 10^{-8}$ | (Torr liters $s^{-1}$) | molecular flow |

The nomographic evaluation of the leakage from atmospheric pressure in an evacuated vessel and the pumping requirements to maintain a certain pressure, are published by Roth (1970, 1990, p. 142).

*b. Leakage from sources within the vacuum system (virtual leaks subscript Lv).*

Refer to Fig. 8.7. The maximum flow rate $Q_{Lv}$ from a trapped volume enclosed in a pumped-down vessel is

$$Q_{Lv\,max.} = p_a V_2/(et) \qquad \text{(e.g., mbar liters s}^{-1}) \qquad (8.12)$$

where $p_a$ (mbar) is the pressure of the trapped gas, $V_2$ (liters), the volume within which the gas is trapped, $e$ = 2.7183 base to natural logarithms, $t$(s), the time subsequent to the start of pumping down the volume $V_1$.

The maximum pressure $p_{max.}$ within the volume $V_1$ produced by the gas escaping from the trapped volume is

$$p_{max.} = Q_{Lv}/S_n \qquad \text{(e.g., mbar)} \qquad (8.13)$$

$$= p_a V_2/(S_n te) \qquad (8.13a)$$

where $S_n$ (liters $s^{-1}$) is the net pumping speed at the port of the vessel and the other symbols are expressed with the same units as in Eq. 8.12.

Figure 8.8 illustrates some trapped volumes usually encountered in vacuum systems.

For more information concerning calculation of the gas load due to the leakage of gas from trapped volumes, see Santeler et al. (1966, pp. 175–180) and Edwards (1979). Calculation of the gas-flow rate owing to leaks through cracks or pinholes in a miner insulated cable exposed to vacuum is studied by Lawson (1966).

## Table 8.9

Rate of Evaporation $\dot{W}$ (g cm$^{-2}$ s$^{-1}$) for Metals and Carbon Corresponding to Vapor Pressures ($\mu$m Hg) and Temperatures $T$ (°C)[a]

| Metal | Data temperature range | $T$ (°C) and $\dot{W}$ | $p_E$($\mu$m Hg) $10^{-2}$ | $10^{-1}$ | 1 | 10 | 100 | 1000 |
|---|---|---|---|---|---|---|---|---|
| Li | 439–1080 | $T$ | 348 | 399 | 460 | 534 | 623 | 737 |
| | | $\dot{W}$ | $6.17 \times 10^{-8}$ | $5.93 \times 10^{-7}$ | $5.68 \times 10^{-6}$ | $5.41 \times 10^{-5}$ | $5.13 \times 10^{-4}$ | $4.84 \times 10^{-3}$ |
| Na | 264–928 | $T$ | 158 | 195 | 238 | 290 | 355 | 437 |
| | | $\dot{W}$ | $1.35 \times 10^{-7}$ | $1.29 \times 10^{-6}$ | $1.24 \times 10^{-5}$ | $1.18 \times 10^{-4}$ | $1.12 \times 10^{-2}$ | $1.05 \times 10^{-2}$ |
| K | 100–760 | $T$ | 91 | 123 | 162 | 208 | 266 | 341 |
| | | $\dot{W}$ | $1.91 \times 10^{-7}$ | $1.83 \times 10^{-6}$ | $1.75 \times 10^{-5}$ | $1.66 \times 10^{-4}$ | $1.57 \times 10^{-3}$ | $1.47 \times 10^{-2}$ |
| Rb | — | $T$ | 64 | 95 | 133 | 176 | 228 | 300 |
| | | $\dot{W}$ | $2.94 \times 10^{-7}$ | $2.81 \times 10^{-6}$ | $2.68 \times 10^{-5}$ | $2.55 \times 10^{-4}$ | $2.41 \times 10^{-2}$ | $2.22 \times 10^{-2}$ |
| Cs | — | $T$ | 64 | 75 | 110 | 152 | 206 | 277 |
| | | $\dot{W}$ | $3.77 \times 10^{-7}$ | $3.61 \times 10^{-6}$ | $3.44 \times 10^{-5}$ | $3.26 \times 10^{-4}$ | $3.07 \times 10^{-3}$ | $2.87 \times 10^{-2}$ |
| Cu | 969–1606 | $T$ | 942 | 1032 | 1142 | 1272 | 1427 | 1622 |
| | | $\dot{W}$ | $1.33 \times 10^{-7}$ | $1.29 \times 10^{-6}$ | $1.24 \times 10^{-5}$ | $1.18 \times 10^{-4}$ | $1.13 \times 10^{-3}$ | $1.07 \times 10^{-2}$ |
| Ag | 721–1000 | $T$ | 757 | 832 | 922 | 1032 | 1167 | 1337 |
| | | $\dot{W}$ | $1.89 \times 10^{-7}$ | $1.82 \times 10^{-6}$ | $1.75 \times 10^{-5}$ | $1.68 \times 10^{-4}$ | $1.60 \times 10^{-3}$ | $1.51 \times 10^{-2}$ |
| Au | 727–987 | $T$ | 987 | 1082 | 1197 | 1332 | 1507 | 1707 |
| | | $\dot{W}$ | $2.31 \times 10^{-7}$ | $2.26 \times 10^{-6}$ | $2.14 \times 10^{-5}$ | $2.05 \times 10^{-4}$ | $1.94 \times 10^{-3}$ | $1.84 \times 10^{-2}$ |
| Be | 899–1279 | $T$ | 902 | 987 | 1092 | 1212 | 1367 | 1567 |
| | | $\dot{W}$ | $5.11 \times 10^{-8}$ | $4.93 \times 10^{-7}$ | $4.74 \times 10^{-6}$ | $4.55 \times 10^{-5}$ | $4.33 \times 10^{-4}$ | $4.08 \times 10^{-3}$ |
| Mg | 736–1020 | $T$ | 287 | 330 | 382 | 442 | 517 | 612 |
| | | $\dot{W}$ | $1.22 \times 10^{-7}$ | $1.17 \times 10^{-6}$ | $1.12 \times 10^{-5}$ | $1.08 \times 10^{-4}$ | $1.02 \times 10^{-3}$ | $0.97 \times 10^{-2}$ |
| Ca | 527–647 | $T$ | 402 | 452 | 517 | 592 | 687 | 817 |
| | | $\dot{W}$ | $1.42 \times 10^{-7}$ | $1.37 \times 10^{-6}$ | $1.31 \times 10^{-5}$ | $1.26 \times 10^{-4}$ | $1.19 \times 10^{-3}$ | $1.12 \times 10^{-2}$ |
| Sr | — | $T$ | 342 | 394 | 436 | 531 | 623 | 742 |
| | | $\dot{W}$ | $2.20 \times 10^{-7}$ | $2.11 \times 10^{-6}$ | $2.02 \times 10^{-5}$ | $1.93 \times 10^{-4}$ | $1.82 \times 10^{-3}$ | $1.71 \times 10^{-2}$ |

| | | | | | | | | |
|---|---|---|---|---|---|---|---|---|
| Ba | 1060–1138 | $T$ | 417 | 467 | 537 | 617 | 727 | 867 |
| | | $\dot{W}$ | $2.60 \times 10^{-7}$ | $2.51 \times 10^{-6}$ | $2.40 \times 10^{-5}$ | $2.28 \times 10^{-4}$ | $2.16 \times 10^{-3}$ | $2.03 \times 10^{-2}$ |
| Zn | 239–377 | $T$ | 208 | 246 | 290 | 342 | 405 | 485 |
| | | $\dot{W}$ | $2.15 \times 10^{-7}$ | $2.07 \times 10^{-6}$ | $1.99 \times 10^{-5}$ | $1.90 \times 10^{-4}$ | $1.81 \times 10^{-3}$ | $1.71 \times 10^{-2}$ |
| Cd | 200–260 | $T$ | 149 | 182 | 221 | 267 | 321 | 392 |
| | | $\dot{W}$ | $3.01 \times 10^{-7}$ | $2.90 \times 10^{-6}$ | $2.78 \times 10^{-5}$ | $2.66 \times 10^{-4}$ | $2.54 \times 10^{-3}$ | $2.44 \times 10^{-2}$ |
| Hg | — | $T$ | −28 | −8 | 16 | 45 | 81 | 125 |
| | | $\dot{W}$ | $5.28 \times 10^{-7}$ | $5.08 \times 10^{-6}$ | $4.86 \times 10^{-5}$ | $4.63 \times 10^{-4}$ | $4.39 \times 10^{-3}$ | $4.14 \times 10^{-2}$ |
| *B* | — | $T$ | 1687 | 1827 | 1977 | 2157 | 2377 | 2657 |
| | | $\dot{W}$ | $4.33 \times 10^{-8}$ | $4.19 \times 10^{-7}$ | $4.05 \times 10^{-6}$ | $3.89 \times 10^{-5}$ | $3.73 \times 10^{-4}$ | $3.55 \times 10^{-3}$ |
| Al | 1137–1195 | $T$ | 882 | 972 | 1082 | 1207 | 1347 | 1547 |
| | | $\dot{W}$ | $8.92 \times 10^{-8}$ | $8.59 \times 10^{-7}$ | $8.23 \times 10^{-6}$ | $7.88 \times 10^{-5}$ | $7.53 \times 10^{-4}$ | $7.10 \times 10^{-3}$ |
| Sc | — | $T$ | 1058 | 1161 | 1282 | 1423 | 1595 | 1804 |
| Y | — | $T$ | | | | | | |
| La | 1327–1627 | $T$ | 1249 | 1362 | 1494 | 1649 | 1833 | 2056 |
| | | $\dot{W}$ | $1.56 \times 10^{-7}$ | $1.69 \times 10^{-6}$ | $1.62 \times 10^{-5}$ | $1.55 \times 10^{-4}$ | $1.48 \times 10^{-3}$ | $1.40 \times 10^{-2}$ |
| Ce | — | $T$ | 1004 | 1091 | 1190 | 1305 | 1439 | 1599 |
| Nd | 899–1279 | $T$ | 957 | 1062 | 1192 | 1342 | 1537 | 1777 |
| | | $\dot{W}$ | $2.00 \times 10^{-7}$ | $1.92 \times 10^{-6}$ | $1.83 \times 10^{-5}$ | $1.74 \times 10^{-4}$ | $1.65 \times 10^{-3}$ | $1.55 \times 10^{-2}$ |
| Ga | 957–1245 | $T$ | 757 | 824 | 937 | 1057 | 1197 | 1372 |
| | | $\dot{W}$ | $1.52 \times 10^{-7}$ | $1.46 \times 10^{-6}$ | $1.40 \times 10^{-5}$ | $1.34 \times 10^{-4}$ | $1.27 \times 10^{-3}$ | $1.22 \times 10^{-2}$ |
| In | 727–1075 | $T$ | 670 | 747 | 837 | 947 | 1077 | 1242 |
| | | $\dot{W}$ | $2.04 \times 10^{-7}$ | $1.96 \times 10^{-6}$ | $1.88 \times 10^{-5}$ | $1.79 \times 10^{-4}$ | $1.70 \times 10^{-3}$ | $1.61 \times 10^{-2}$ |
| Tl | — | $T$ | 412 | 468 | 535 | 615 | 713 | 837 |
| | | $\dot{W}$ | $3.19 \times 10^{-7}$ | $3.06 \times 10^{-6}$ | $2.93 \times 10^{-5}$ | $2.80 \times 10^{-4}$ | $2.66 \times 10^{-3}$ | $2.50 \times 10^{-2}$ |
| C | 2084–2597 | $T$ | 1977 | 2107 | 2247 | 2427 | 2627 | 2867 |
| | | $\dot{W}$ | $4.27 \times 10^{-8}$ | $4.14 \times 10^{-7}$ | $4.03 \times 10^{-6}$ | $3.89 \times 10^{-5}$ | $3.76 \times 10^{-4}$ | $3.61 \times 10^{-3}$ |
| Si | — | $T$ | 1177 | 1282 | 1357 | 1547 | 1717 | 1927 |
| | | $\dot{W}$ | $8.12 \times 10^{-8}$ | $7.84 \times 10^{-7}$ | $7.54 \times 10^{-6}$ | $7.24 \times 10^{-5}$ | $6.93 \times 10^{-4}$ | $6.59 \times 10^{-2}$ |
| Ti | 1111–1323 | $T$ | 1321 | 1431 | 1558 | 1703 | 1877 | 2083 |
| | | $\dot{W}$ | $1.01 \times 10^{-7}$ | $0.98 \times 10^{-6}$ | $0.94 \times 10^{-5}$ | $0.90 \times 10^{-4}$ | $0.86 \times 10^{-3}$ | $0.82 \times 10^{-2}$ |

*continues*

**Table 8.9** *Continued*

| Metal | Data Temperature Range | $T$ (°C) and $\dot{W}$ | $p_E(\mu m\ Hg)$ | | | | | |
|---|---|---|---|---|---|---|---|---|
| | | | $10^{-2}$ | $10^{-1}$ | 1 | 10 | 100 | 1000 |
| Zr | 1676–1781 | $T$ | 1837 | 2002 | 2187 | 2397 | 2647 | 2977 |
| | | $\dot{W}$ | $1.21 \times 10^{-7}$ | $1.17 \times 10^{-5}$ | $1.12 \times 10^{-5}$ | $1.08 \times 10^{-4}$ | $1.03 \times 10^{-3}$ | $0.98 \times 10^{-2}$ |
| Th | — | $T$ | 1686 | 1831 | 1999 | 2196 | 2431 | 2715 |
| | | $\dot{W}$ | $2.01 \times 10^{-7}$ | $1.94 \times 10^{-6}$ | $1.86 \times 10^{-5}$ | $1.79 \times 10^{-4}$ | $1.71 \times 10^{-3}$ | $1.63 \times 10^{-2}$ |
| Ge | 1237–1612 | $T$ | 1037 | 1142 | 1262 | 1407 | 1582 | 1797 |
| | | $\dot{W}$ | $1.37 \times 10^{-7}$ | $1.32 \times 10^{-6}$ | $1.27 \times 10^{-5}$ | $1.21 \times 10^{-5}$ | $1.15 \times 10^{-3}$ | $1.09 \times 10^{-2}$ |
| Sn | 1151–1415 | $T$ | 882 | 977 | 1092 | 1227 | 1397 | 1612 |
| | | $\dot{W}$ | $1.87 \times 10^{-7}$ | $1.80 \times 10^{-6}$ | $1.72 \times 10^{-5}$ | $1.64 \times 10^{-4}$ | $1.56 \times 10^{-3}$ | $1.46 \times 10^{-2}$ |
| Pb | — | $T$ | 487 | 551 | 627 | 719 | 832 | 977 |
| | | $\dot{W}$ | $3.05 \times 10^{-7}$ | $2.93 \times 10^{-6}$ | $2.80 \times 10^{-5}$ | $2.67 \times 10^{-4}$ | $2.53 \times 10^{-3}$ | $2.38 \times 10^{-2}$ |
| V | 1389–1609 | $T$ | 1432 | 1551 | 1687 | 1847 | 2937 | 2287 |
| | | $\dot{W}$ | $1.01 \times 10^{-7}$ | $0.98 \times 10^{-6}$ | $0.94 \times 10^{-5}$ | $0.90 \times 10^{-4}$ | $0.87 \times 10^{-3}$ | $0.82 \times 10^{-2}$ |
| Cb | — | $T$ | 2194 | 2355 | 2539 | — | — | — |
| | | $\dot{W}$ | $1.16 \times 10^{-7}$ | $1.08 \times 10^{-6}$ | $1.06 \times 10^{-5}$ | — | — | — |
| Ta | 1727–2997 | $T$ | 2397 | 2587 | 2807 | 3067 | 3372 | 3737 |
| | | $\dot{W}$ | $1.52 \times 10^{-7}$ | $1.47 \times 10^{-6}$ | $1.41 \times 10^{-5}$ | $1.36 \times 10^{-4}$ | $1.30 \times 10^{-3}$ | $1.24 \times 10^{-2}$ |
| $P_4$ | — | $T$ | 107 | 130 | 157 | 187 | 222 | 262 |
| | | $\dot{W}$ | $3.3 \times 10^{-7}$ | $3.24 \times 10^{-6}$ | $3.13 \times 10^{-5}$ | $3.03 \times 10^{-4}$ | $2.92 \times 10^{-3}$ | $2.81 \times 10^{-2}$ |
| $Sb_2$ | — | $T$ | 382 | 427 | 477 | 542 | 617 | 757 |
| | | $\dot{W}$ | $2.52 \times 10^{-7}$ | $2.43 \times 10^{-6}$ | $2.35 \times 10^{-5}$ | $2.26 \times 10^{-4}$ | $2.16 \times 10^{-3}$ | $2.01 \times 10^{-2}$ |
| Bi | 409–497 | $T$ | 450 | 508 | 578 | 661 | 762 | 892 |
| | | $\dot{W}$ | $3.14 \times 10^{-7}$ | $3.02 \times 10^{-6}$ | $2.89 \times 10^{-5}$ | $2.76 \times 10^{-4}$ | $2.62 \times 10^{-3}$ | $2.47 \times 10^{-2}$ |
| Cr | 889–1228 | $T$ | 1062 | 1162 | 1267 | 1392 | 1557 | 1737 |
| | | $\dot{W}$ | $1.15 \times 10^{-7}$ | $1.11 \times 10^{-6}$ | $1.07 \times 10^{-5}$ | $1.03 \times 10^{-4}$ | $0.98 \times 10^{-3}$ | $0.94 \times 10^{-2}$ |
| Mo | 1797–2231 | $T$ | 1987 | 2167 | 2377 | 2627 | 2927 | 3297 |
| | | $\dot{W}$ | $1.20 \times 10^{-7}$ | $1.16 \times 10^{-6}$ | $1.11 \times 10^{-5}$ | $1.06 \times 10^{-4}$ | $1.01 \times 10^{-3}$ | $0.95 \times 10^{-2}$ |
| W | — | $T$ | 2547 | 2757 | 3007 | 3297 | 3647 | — |
| | | $\dot{W}$ | $1.49 \times 10^{-7}$ | $1.44 \times 10^{-6}$ | $1.38 \times 10^{-5}$ | $1.32 \times 10^{-4}$ | $1.26 \times 10^{-3}$ | — |

| | | | | | | | | |
|---|---|---|---|---|---|---|---|---|
| U | 1357–1697 | $T$ | 1442 | 1582 | 1737 | 1927 | 2157 | 2447 |
| | | $\dot{W}$ | $2.17 \times 10^{-7}$ | $2.09 \times 10^{-6}$ | $2.01 \times 10^{-5}$ | $1.92 \times 10^{-4}$ | $1.83 \times 10^{-3}$ | $1.73 \times 10^{-2}$ |
| Se | — | $T$ | 144 | 167 | 197 | 232 | 277 | 347 |
| | | $\dot{W}$ | $2.54 \times 10^{-7}$ | $2.47 \times 10^{-6}$ | $2.39 \times 10^{-5}$ | $2.31 \times 10^{-4}$ | $2.21 \times 10^{-3}$ | $2.08 \times 10^{-2}$ |
| $Te_2$ | — | $T$ | 261 | 296 | 336 | 383 | 438 | 520 |
| | | $\dot{W}$ | $4.03 \times 10^{-7}$ | $3.91 \times 10^{-6}$ | $3.78 \times 10^{-5}$ | $3.64 \times 10^{-4}$ | $3.50 \times 10^{-3}$ | $3.31 \times 10^{-2}$ |
| Po | 438–745 | $T$ | 187 | 220 | 263 | 314 | 382 | 472 |
| | | $\dot{W}$ | $3.94 \times 10^{-7}$ | $3.81 \times 10^{-6}$ | $3.65 \times 10^{-5}$ | $3.49 \times 10^{-4}$ | $3.30 \times 10^{-3}$ | $3.10 \times 10^{-2}$ |
| Mn | — | $T$ | 697 | 767 | 852 | 947 | 1067 | 1227 |
| | | $\dot{W}$ | $1.39 \times 10^{-7}$ | $1.34 \times 10^{-6}$ | $1.29 \times 10^{-5}$ | $1.24 \times 10^{-4}$ | $1.18 \times 10^{-3}$ | $1.12 \times 10^{-2}$ |
| Re | — | $T$ | 2367 | 2557 | 2787 | 3057 | 3397 | — |
| | | $\dot{W}$ | $1.55 \times 10^{-7}$ | $1.50 \times 10^{-6}$ | $1.44 \times 10^{-5}$ | $1.38 \times 10^{-4}$ | $1.31 \times 10^{-3}$ | — |
| Fe | 1092–1246 | $T$ | 1107 | 1207 | 1322 | 1467 | 1637 | 1847 |
| | | $\dot{W}$ | $1.17 \times 10^{-4}$ | $1.13 \times 10^{-6}$ | $1.09 \times 10^{-5}$ | $1.05 \times 10^{-4}$ | $0.99 \times 10^{-3}$ | $0.95 \times 10^{-2}$ |
| Co | 1090–1249 | $T$ | 1162 | 1262 | 1377 | 1517 | 1697 | 1907 |
| | | $\dot{W}$ | $1.18 \times 10^{-7}$ | $1.14 \times 10^{-6}$ | $1.10 \times 10^{-5}$ | $1.06 \times 10^{-4}$ | $1.01 \times 10^{-3}$ | $0.96 \times 10^{-2}$ |
| Ni | 1034–1310 | $T$ | 1142 | 1247 | 1357 | 1497 | 1667 | 1877 |
| | | $\dot{W}$ | $1.19 \times 10^{-7}$ | $1.15 \times 10^{-6}$ | $1.11 \times 10^{-5}$ | $1.06 \times 10^{-4}$ | $1.01 \times 10^{-3}$ | $0.96 \times 10^{-2}$ |
| Ru | — | $T$ | 1913 | 2058 | 2230 | 2431 | 2666 | 2946 |
| | | $\dot{W}$ | $1.26 \times 10^{-7}$ | $1.22 \times 10^{-6}$ | $1.18 \times 10^{-5}$ | $1.13 \times 10^{-4}$ | $1.08 \times 10^{-3}$ | $1.04 \times 10^{-2}$ |
| Rh | — | $T$ | 1587 | 1707 | 1857 | 2027 | 2247 | 2527 |
| | | $\dot{W}$ | $1.37 \times 10^{-7}$ | $1.33 \times 10^{-6}$ | $1.28 \times 10^{-5}$ | $1.23 \times 10^{-4}$ | $1.18 \times 10^{-3}$ | $1.12 \times 10^{-2}$ |
| Pd | — | $T$ | 1157 | 1262 | 1387 | 1547 | 1727 | 1967 |
| | | $\dot{W}$ | $1.59 \times 10^{-7}$ | $1.54 \times 10^{-6}$ | $1.48 \times 10^{-5}$ | $1.41 \times 10^{-4}$ | $1.35 \times 10^{-3}$ | $1.27 \times 10^{-2}$ |
| Os | — | $T$ | 2101 | 2264 | 2451 | 2667 | 2920 | 3221 |
| | | $\dot{W}$ | $1.65 \times 10^{-7}$ | $1.60 \times 10^{-6}$ | $1.54 \times 10^{-5}$ | $1.48 \times 10^{-4}$ | $1.42 \times 10^{-3}$ | $1.36 \times 10^{-2}$ |
| Ir | — | $T$ | 1797 | 1947 | 2107 | 2307 | 2527 | 2827 |
| | | $\dot{W}$ | $1.78 \times 10^{-7}$ | $1.72 \times 10^{-6}$ | $1.66 \times 10^{-5}$ | $1.60 \times 10^{-4}$ | $1.53 \times 10^{-3}$ | $1.46 \times 10^{-2}$ |
| Pt | — | $T$ | 1602 | 1742 | 1907 | 2077 | 2317 | 2587 |
| | | $\dot{W}$ | $1.88 \times 10^{-7}$ | $1.82 \times 10^{-6}$ | $1.75 \times 10^{-5}$ | $1.68 \times 10^{-4}$ | $1.60 \times 10^{-3}$ | $1.52 \times 10^{-2}$ |

[a]After Dushman (1966, pp. 696–699).

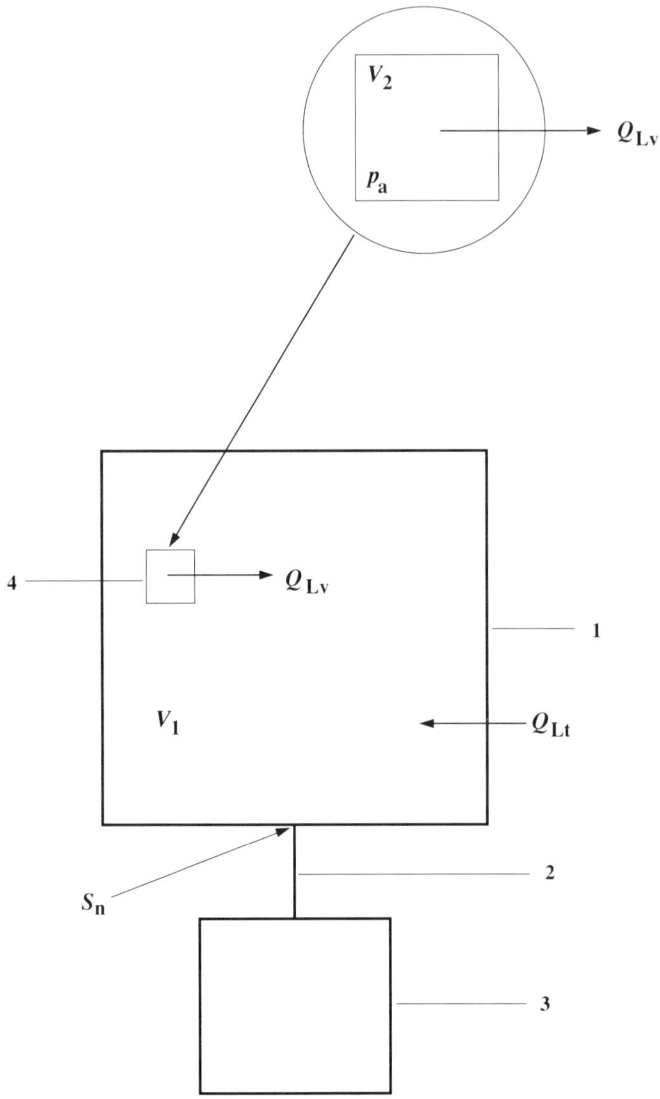


**Figure 8.7** Occurrence of leaks in a vacuum system: 1—vessel, 2 pipe, 3—vacuum pump, 4—source of virtual leak, $S_n$—pumping speed at the port of the vessel (1), $V_1$—volume of the vessel (1), $V_2$—volume of the trapped gas, $Q_{Lt}$—flow rate of gas from true leaks, $Q_{Lv}$—flow rate of gas from virtual leaks.

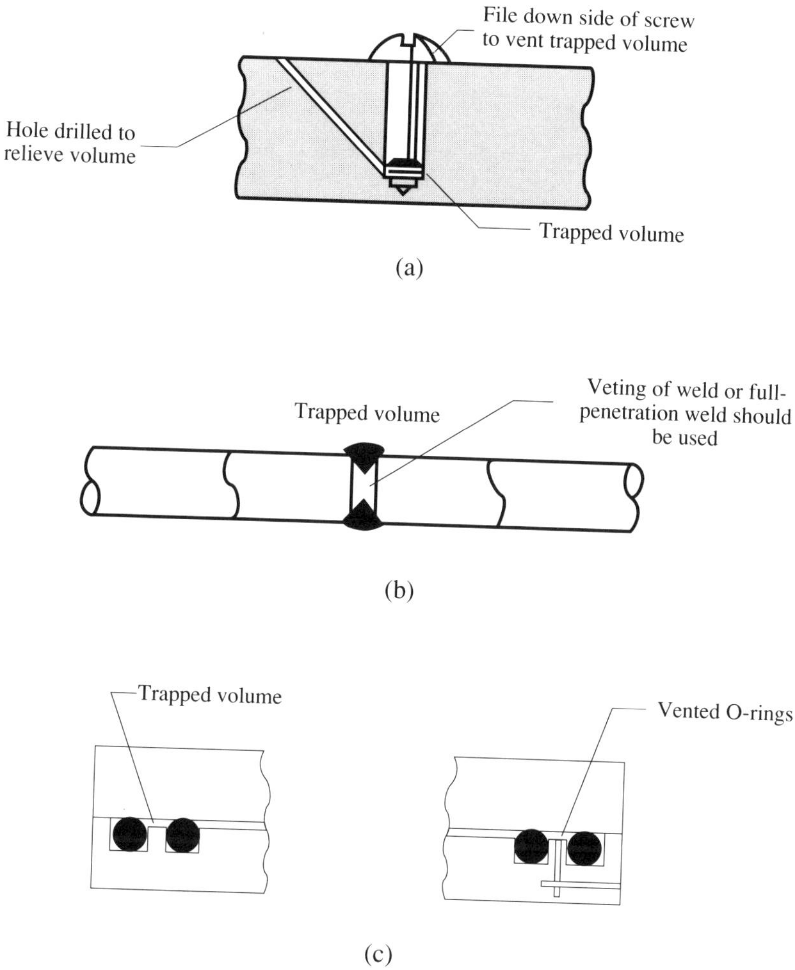


**Figure 8.8** Schematic of some trapped volumes encountered in vacuum systems and suggested solutions for releasing the trapped gas: (a) bottom of a blind trapped screw, (b) two welds in series, (c) O-rings in series. [After Santeler et al. (1966).]

### 8.2.4 Permeation

The gas load $Q_{\mathrm{Per}}$ due to permeation (denoted by subscript Per) of gas through solid slabs is

$$Q_{\mathrm{Per}} = K_{\mathrm{Per}} A \Delta_{\mathrm{p}} L^{-1} \qquad (\mathrm{atm\ cm^3\ s^{-1}}) \tag{8.14}$$

where $K_{\mathrm{Per}}$ ($\mathrm{cm^3}$ at STP $\mathrm{cm^{-2}\ s^{-1}}$ for 1-atm pressure gradient per millimeter of thickness) is the permeability coefficient, $A$ ($\mathrm{cm^2}$) is the surface of the material, $\Delta p$ (atm) is the pressure gradient across the thickness of the material, and $L$(cm) is the thickness of the material.

To convert $K_{\mathrm{Per}}$ ($\mathrm{cm^3}$ at STP $\mathrm{cm^{-2}\ s^{-1}}$ for 1 atm per 1-mm thickness) to

$K_{\mathrm{Per}}(\mathrm{m^2\ s^{-1}})$ when pressure is in pascals, multiply by $10^{-5}$
$K_{\mathrm{Per}}$ ($\mathrm{cm^2\ s^{-1}}$) when pressure is in mbar or Torr, multiply by $10^{-4}$

**Table 8.10**
Permeability Coefficient $K_{\mathrm{Per}}$ for Some Polymers

| Gas / Material | $K_{\mathrm{Per}}$ $\mathrm{cm^3}$ (STP) $\mathrm{cm^{-2}\ s^{-1}}$ for 1 atm $\mathrm{mm^{-1}}$ (thickness) at $T = 23°\mathrm{C}$ | | | | | |
|---|---|---|---|---|---|---|
| | $CO_2$ | $H_2$ | He | $N_2$ | $O_2$ | Ref. |
| Buna | $9.18 \times 10^{-5}$ | — | — | $4.69 \times 10^{-7}$ | — | Rogers et al. (1956) |
| Butyl rubber | $3.81 \times 10^{-6}$ | — | — | $1.76 \times 10^{-8}$ | $9.67 \times 10^{-8}$ | |
| Cellulose acetate | $5.76 \times 10^{-8}$ | — | — | $2.05 \times 10^{-8}$ | — | |
| Kel-F[a] | — | — | — | $9.64 \times 10^{-8}$ | $4.16 \times 10^{-8}$ | |
| Methyl rubber | $5.57 \times 10^{-7}$ | — | — | $3.52 \times 10^{-8}$ | $1.56 \times 10^{-7}$ | |
| Natural rubber | $9.67 \times 10^{-6}$ | — | — | $5.96 \times 10^{-7}$ | $1.73 \times 10^{-6}$ | |
| Neoprene | $1.86 \times 10^{-6}$ | — | — | $8.79 \times 10^{-8}$ | — | |
| Perbunan | $2.30 \times 10^{-6}$ | — | — | $7.86 \times 10^{-8}$ | — | |
| Plyofilm | $1.93 \times 10^{-8}$ | — | — | $5.92 \times 10^{-10}$ | — | |
| Polyamide (nylon) | $1.18 \times 10^{-8}$ | — | — | $7.42 \times 10^{-10}$ | — | |
| Polyester (mylar) | $1.13 \times 10^{-8}$ | — | — | $3.71 \times 10^{-10}$ | $1.63 \times 10^{-9}$ | |
| Saran[b] | $2.15 \times 10^{-9}$ | — | — | $7.00 \times 10^{-11}$ | $3.95 \times 10^{-10}$ | |
| Viton A | — | $2.2 \times 10^{-7}$ | $8.2 \times 10^{-7}$ | — | — | Weston (1975) |

[a]Policychloroethylene
[b]Polyvinylidene chloride.

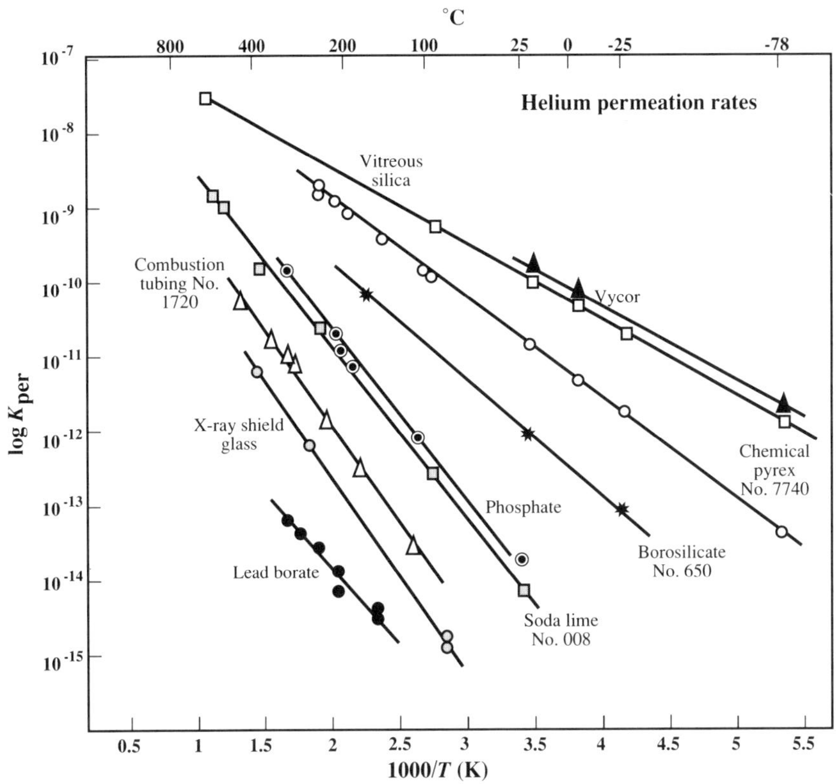


**Figure 8.9** Permeation rate $K_{\mathrm{Per}}$ of helium through various glasses. [After Norton (1962).]

To convert $Q_{\mathrm{Per}}$ from STP conditions to a temperature $T_{\mathrm{K}}$, multiply by $(T_{\mathrm{K}}/273\ \mathrm{K})$.

Hydrogen easily permeates stainless steel, but at a rate 10–100 slower than that in chrome steel (Norton, 1962).

Metals are permeated by $H_2$ at a rate increasing with the square root of the driving pressure (Norton, 1962).

*Rare gases do not permeate metals.*

The values of the permeability coefficient $K_{\mathrm{Per}}$ for some polymers are listed in Table 8.10.

The helium permeation rate through various glasses, and the atmospheric gas accumulation in silica and other glass bulbs is shown in Figs. 8.9–8.11.

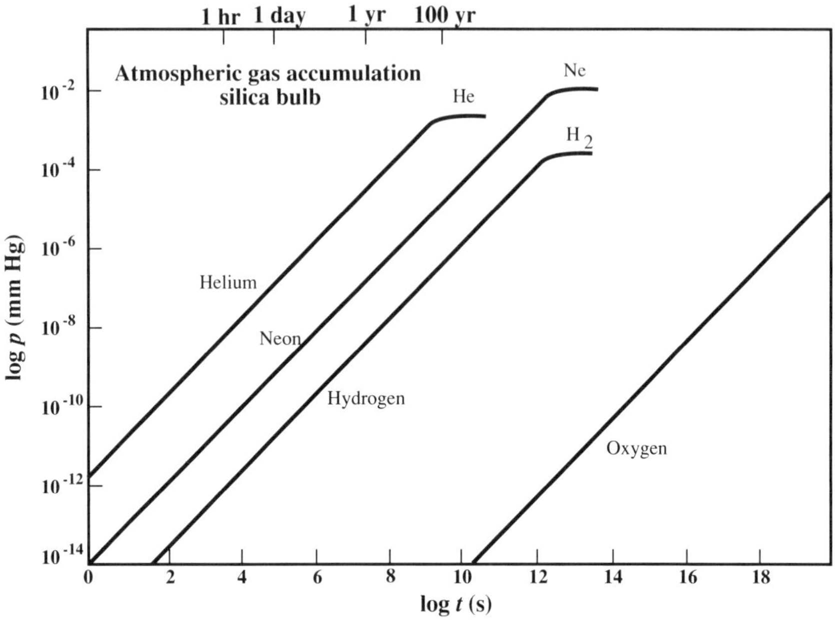


**Figure 8.10** Atmospheric gas accumulation at 25°C in a silica bulb of 330 cm$^3$, 1 mm wall thickness, 100 cm$^2$ area. [After Norton (1962).]

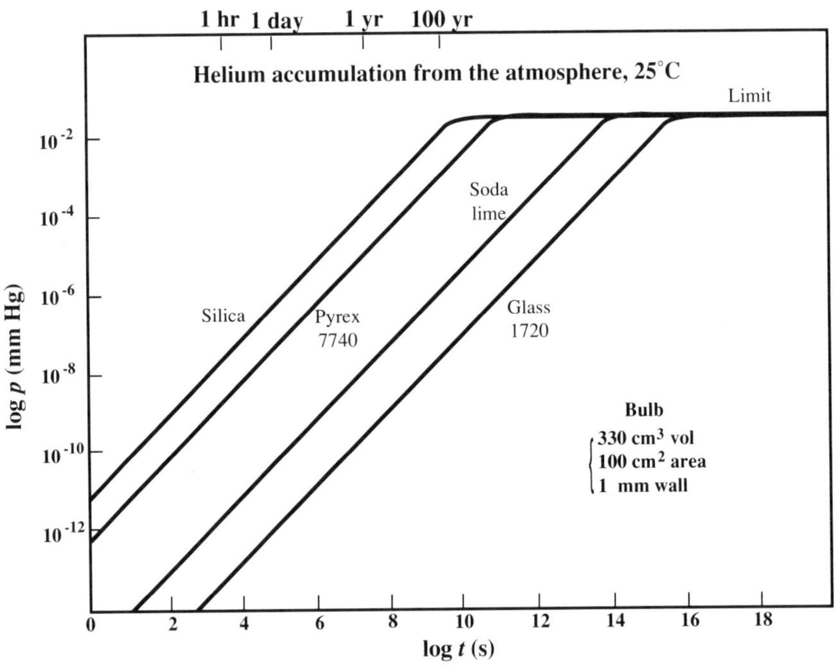


**Figure 8.11** Helium accumulation from the atmosphere in bulbs of various glasses. [After Norton (1962).]

## Exercises

### Exercise 8.1

Calculate the mass in grams corresponding to an amount of $10^{-7}$ mbar of He at (a) 0°C and (b) 23°C.

### Solution 8.1

(a) Use Eq. 8.4 written as

$$10^{-7} \text{ mbar} = 10^{-7} \times 4.406 \times 10^{-5} M \qquad (\text{g})$$

Substitute

$$M_{\text{He}} = 4.0 \qquad (\text{see Table 1.2})$$

Thus

$$10^{-7} \text{ mbar liters (at 0°C)} = 10^{-7} \times 4.406 \times 10^{-5} \times 4.0 = 1.76 \times 10^{-11} \text{ g}$$

(b) To convert the amount (g) at 0°C (273 K) to 23°C (273 + 23 = 296 K), multiply by 273 K/296 K. Hence

$$10^{-7} \text{ mbar liters (at 23°C)} = 1.76 \times 10^{-11} \text{ g} \times 273 \text{ K}/296 \text{ K}$$
$$= 1.63 \times 10^{-11} \text{ g}$$

### Exercise 8.2

A turbomolecular pump of capacity 50 liters $s^{-1}$ is connected to the port of a vessel accommodating a cavity. The cavity contains air at atmospheric pressure, trapped within a volume of 1 $cm^3$. Calculate the maximum load and the pressure in the vessel owing to the virtual gas leak after the first hour of pumpdown.

### Solution 8.2

Use Eqs. 8.12 and 8.13:

$$Q_{\text{Lv max.}} = p_a V_2/(et)$$
$$p_{\text{max.}} = Q_{\text{Lv}}/S_n$$

Substitute

$p_a = 760$ Torr
$V_2 = 1 \text{ cm}^3 = 10^{-3}$ liters
$S_n = 50$ liters $s^{-1}$
$t = 1 \text{ h} = 3600$ s
$e = 2.7183$

Thus

$$Q_{\text{Lv}} = 760 \text{ Torr} \times 10^{-3} \text{ liters}/(2.7183 \times 3600 \text{ s})$$
$$= 7.8 \times 10^{-5} \text{ Torr liters s}^{-1}$$
$$p_a = 7.8 \times 10^{-5} \text{ Torr liters s}^{-1}/50 \text{ liters s}^{-1} = 1.5 \times 10^{-6} \text{ Torr}$$

**Exercise 8.3**

Owing to a leak in a 100-liter vacuum vessel separated from the pump, the pressure in the vessel increases from $5 \times 10^{-4}$ to $5 \times 10^{-2}$ mbar in 120 s. At the end of this time interval, the vessel is pumped down by a pump of 50 liters $s^{-1}$, directly connected to its port. What is the pressure in the pumped-down vessel, assuming that the temperature of the vessel remains constant?

**Solution 8.3**

Use Eqs. 4.6 and 4.38a written as

$$Q_{Lt} = (\Delta p/\Delta t)V$$

$$p = Q_{Lt}/S_n$$

Substitute

$$p = (5 \times 10^{-2}\ \text{mbar} - 5 \times 10^{-4}\ \text{mbar}) = 4.95 \times 10^{-2}\ \text{mbar}$$
$$t = 120\ \text{s}$$
$$S_n = 50\ \text{liters s}^{-1} = \text{net (effective) pumping speed at the port of the vessel}$$
$$V = 100\ \text{liters}$$

Hence

$$Q_{Lt} = \left(4.95 \times 10^{-2}\ \text{mbar}/120\ \text{s}\right) \times 100\ \text{liters}$$

$$= 4.125 \times 10^{-2}\ \text{mbar liters s}^{-1}$$

$$p = 4.125 \times 10^{-2}\ \text{mbar liters s}^{-1}/50\ \text{liters s}^{-1} = 8.25 \times 10^{-4}\ \text{mbar}$$

## References

Blears, J., Greer, E. I., and Nightingale, I. (1960), *Transactions of 1st International Vacuum Congress*, p. 473, Pergamon Press, Oxford.

CRC (1986–1987), *Handbook of Chemistry and Physics*, 67th ed. R. C. Weast (ed.), CRC Press, Boca Raton, Fla.

Dayton, B. B. (1959), *Transactions of 5th American Vacuum Symposium*, p. 101, Pergamon Press, Oxford.

Dayton, B. B. (1960), *Transactions of 6th American Vacuum Symposium*, p. 101, Pergamon Press, Oxford.

Dayton, B. B. (1962), *Transactions of 2nd International Vacuum Congress*, p. 42, Pergamon Press, Oxford.

Dayton, B. B. (1963), *Transactions 9th American Vacuum Symposium*, p. 293, Macmillan, New York.

Delafosse, J., and Mongodin, G. (1961), *Le Vide* 3, 22.

Dushman, S. (1966), in *Scientific Foundations of Vacuum Technique*, J. M. Lafferty (ed.), Wiley, New York.

Edwards, D., Jr. (1979), *Vacuum*, *29*, 169.
Geller, R. (1958), *Le Vide*, *13*, 71.
Honig, R. E., and Kramer, D. A. (1960), © *RCA Rev.*, *21*, 360.
Honig, R. E., and Kramer, D. A. (1969), © *RCA Rev.*, *30*, 285.
Lawson, J. D. (1966), *J. Sci. Instrum.*, *43*, 565.
Norton, F. J. (1962), *Transactions of AVS National Vacuum Symposium* (1961), p. 8, Pergamon Press, New York.
Power, B. D., and Crawley, D. J. (1960), *Advances in Vacuum Science and Technology*, Vol. I, p. 206, Pergamon Press, Oxford.
Rogers, C., Meyer, J. A., Stannet, V., and Szwarc, M. (1956), *J. Tech. Assoc. Pulp & Paper Industry (TAPPI)*, *39*, 11, 737–747.
Rosenblum, St. S. (1986), *J. Vac. Sci. Technol. A4*, *1*, 107.
Roth, A. (1970), Report IA-1220 (Israel Atomic Energy Commission).
Roth, A. (1990), *Vacuum Technology*, North-Holland, Amsterdam.
Santeler, D. J., Holkeboer, D. H., Jones, D. W., and Pagano, F. (1966), *Vacuum Technology and Space Simulation*, National Aeronautics and Space Administration, NASA SP-105, Washington, D.C.
Schram, A. (1963), *Le Vide*, *103*, 55.
Weston, G. F. (1975), *Vacuum*, *25*, 469.

# 9

# Vacuum Pumps

## 9.1 *Survey of Various Types of Vacuum Pumps*

Vacuum pumps are used to remove gas molecules in the gas phase from a gas-filled volume and to maintain a required degree of gas rarefaction in that volume. No single pump can evacuate gas from atmospheric pressure down to the ultra-high-vacuum range. To extend their lower-pressure range, pumps are associated in systems in order to increase their lower-range capabilities.

The pressure ranges of vacuum pumps are illustrated in Fig. 9.1.

Figure 9.2 shows a qualitative illustration of the pressure ranges in which the use of various types of vacuum pumps is most economical (i.e., \$/liters $s^{-1}$).

The performance characteristics of some vacuum pumps are illustrated in Figs. 9.3–9.6.

## 9.2 *Positive-Displacement Pumps*

### 9.2.1 Compression Ratio

$$C_r = p_e/(p_v + p_g) \qquad \text{(dimensionless)} \qquad (9.1)$$

where $C_r$ is the compression ratio; $p_e$ (e.g., mbar) is the pressure required to raise the exhaust valve of the pump against the combined pressures of the spring, oil, and air acting on that valve; $p_v$ (e.g., mbar) is the partial pressure of vapor at the pump inlet; and $p_g$ (e.g., mbar) is the partial pressure of the permanent gas at the port of the pump.

The value of $p_e$ is quoted by the manufacturer in the data sheet of the pump. Usually $p_e$ = 1300–1400 mbar.

### 9.2.2 Simultaneous Pumping of Gas and Vapor

The maximum partial pressure of vapor (water) $p_{v\,max}$ that can be pumped without causing condensation in the pump is as follows.

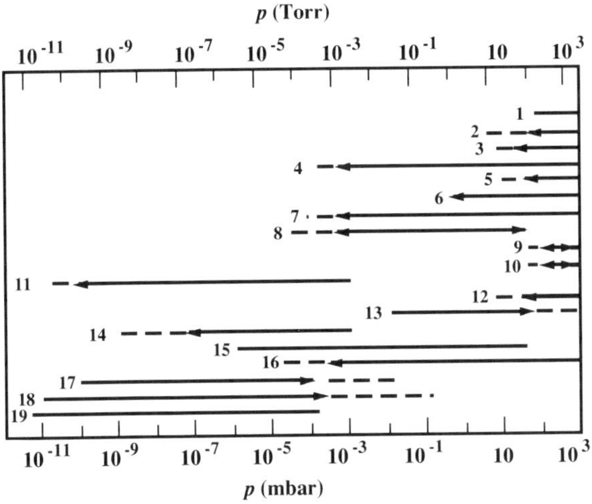


**Figure 9.1** Pressure range of vacuum pumps: 1—piston, 2—diaphragm, 3—water ring, 4—sliding vane, 5—multiple vane, 6—rotary piston, 7—rotary plunger, 8—Roots, 9—turbine, 10—gaseous ring, 11—turbomolecular, 12—liquid jet, 13—gas jet, 14—vapor jet, 15—diffusion ejector, 16—adsorption, 17—sublimation, 18—sputter-ion, 19—cryogenic.

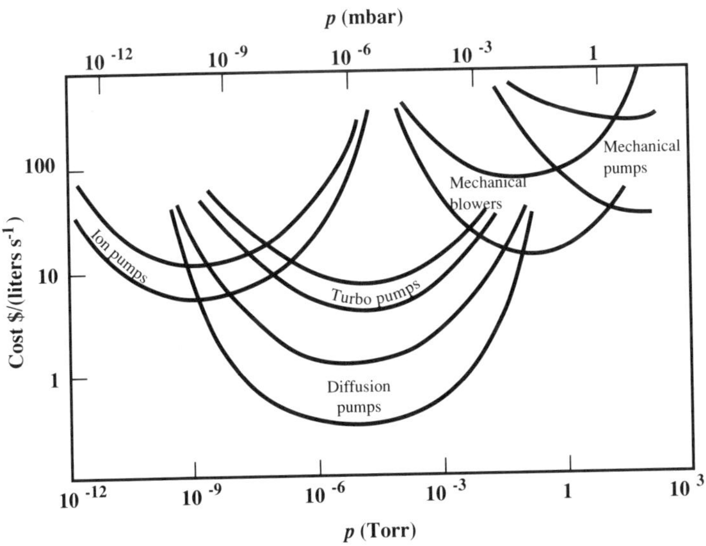


**Figure 9.2** Qualitative illustration of the pressure ranges in which the use of various types of vacuum pumps is most economical. [After Hablanian (1979).]

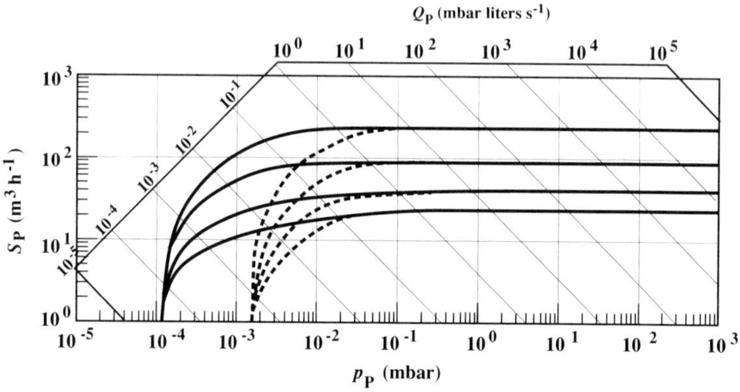


**Figure 9.3** Performance characteristic of double-stage rotary-vane vacuum pumps, without gas ballast (continuous line) and with gas ballast (interrupted line). Ultimate pressure of the pumps without gas ballast $1.2 \times 10^{-1}$ mbar, with gas ballast $1.2 \times 10^{-3}$ mbar. Nominal pumping speeds: (a) 250 $m^3\ h^{-1}$, (b) 110 $m^3\ h^{-1}$, (c) 50 $m^3\ h^{-1}$.

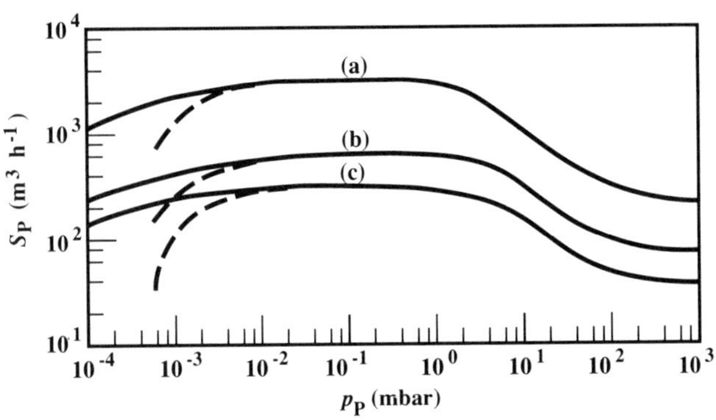


**Figure 9.4** Performance characteristics of Roots pumps. Continuous line, the gas ballast of the backing pump shut off. Interrupted line, the gas ballast of the backing pump open. Ultimate pressure $5.0 \times 10^{-2}$ mbar (single-stage backing pump), $1.0 \times 10^{-4}$ mbar (double-stage backing pump). Nominal pumping speeds: (a) 1500 $m^3\ h^{-1}$, (b) 600 $m^3\ h^{-1}$, (c) 250 $m^3\ h^{-1}$.

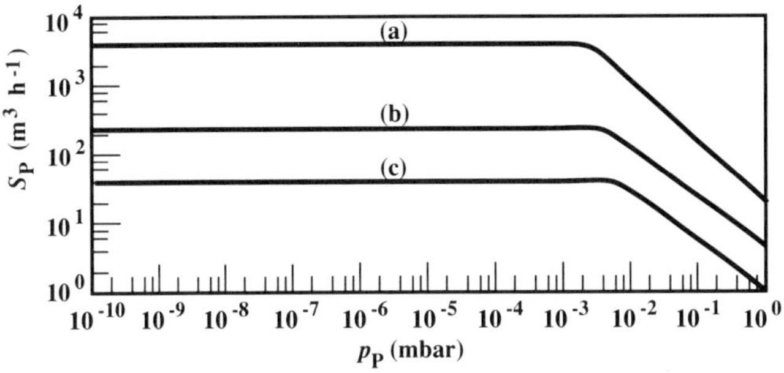


**Figure 9.5** Performance characteristics for air of turbomolecular pumps. Nominal pumping speeds: (a) 4000 $m^3\ h^{-1}$, (b) 125 $m^3\ h^{-1}$, (c) 40 $m^3\ h^{-1}$.

*Rotary pumps without gas ballast.*

$$P_{v\,max.} = p_{sv}p_g/(p_e - p_{sv}) \qquad \text{(e.g., mbar)} \qquad (9.2)$$

where $p_{sv}$ is the pressure of the saturated vapor at the temperature of the pump. The values of $p_{sv}$ as a function of temperature can be found in tables (see, e.g., Table 8.8) and graphs (see, e.g., Fig. 8.6).

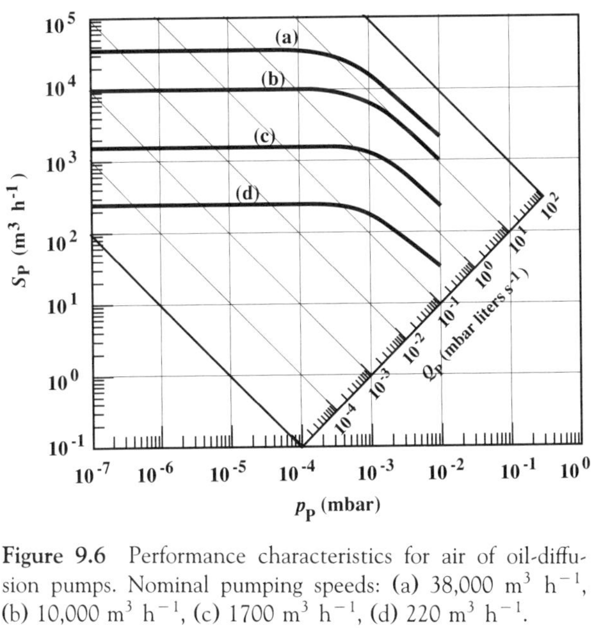


**Figure 9.6** Performance characteristics for air of oil-diffusion pumps. Nominal pumping speeds: (a) 38,000 $m^3\ h^{-1}$, (b) 10,000 $m^3\ h^{-1}$, (c) 1700 $m^3\ h^{-1}$, (d) 220 $m^3\ h^{-1}$.

When the pump exhausts through a filter, the pressure $p_e$ in Eq. 9.2 must be augmented by the pressure difference $\Delta p$ (usually quoted by the manufacturer) across the filter.

*Rotary pumps with gas ballast.*

$$p_{v\,max} = (p_{sv}/S_p)(T_p/T_a) \times [p_g S_P + (p_B - p_p)B]/[(p_e - p_{sv}) - Bp_p/p_{sv}] \quad \text{(e.g., mbar)} \quad (9.3)$$

where, in addition to the notations used in Eqs. 9.1 and 9.2, $S_P$ (e.g., liters $s^{-1}$), is the volume rate of flow (pumping speed) at the port of the pump; $T_P$ (K), the

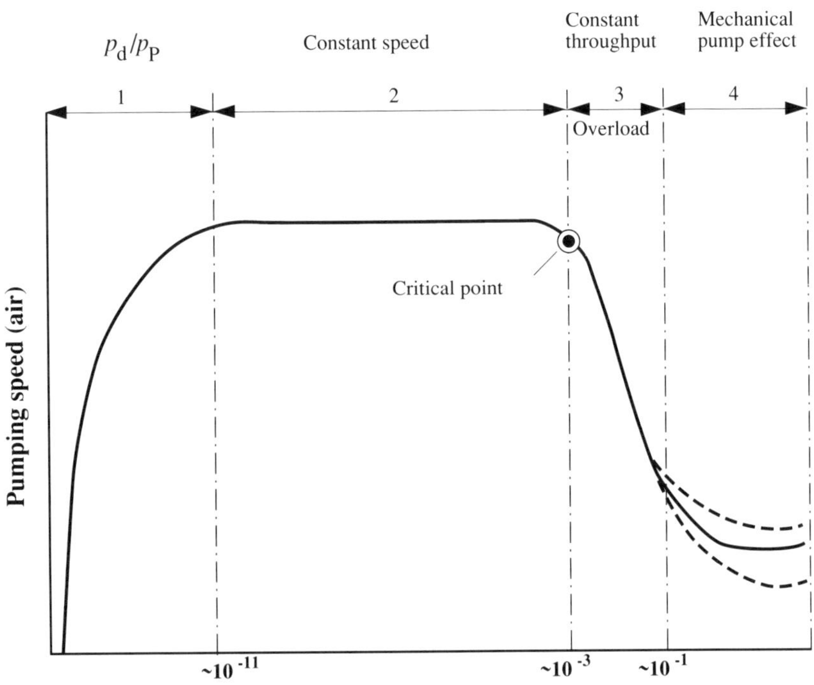


**Figure 9.7** Typical speed–pressure curve of a diffusion pump: $p_a$—discharge pressure, $p_p$—inlet pressure. [After Hablanian (1974).]

pump temperature; $T_a$ (K), the ambient temperature $P_B$ (e.g., mbar), the total pressure of the air ballast; $p_p$ (e.g., mbar), the partial pressure of vapor in the ballast; and $B$ (e.g., liters $s^{-1}$), the gas ballast volume rate of flow.

## 9.3 *Diffusion Pumps*

The plot of the pumping speed of a diffusion pump versus the inlet pressure (see Fig. 9.7) consists of four sections (Hablanian, 1974).

Section 1 (in Fig. 9.7) shows the effect of pumping speed decrease owing to the limit of maximum discharge pressure ($p_d$) inlet pressure ($p_P$) ratio. Helium and hydrogen produce this effect. In section 2, the amount of gas being pumped is lower than the maximum capacity of the pump; hence, the pumping speed remains constant. Section 3, marked "overload," is a *constant-throughput* sec-

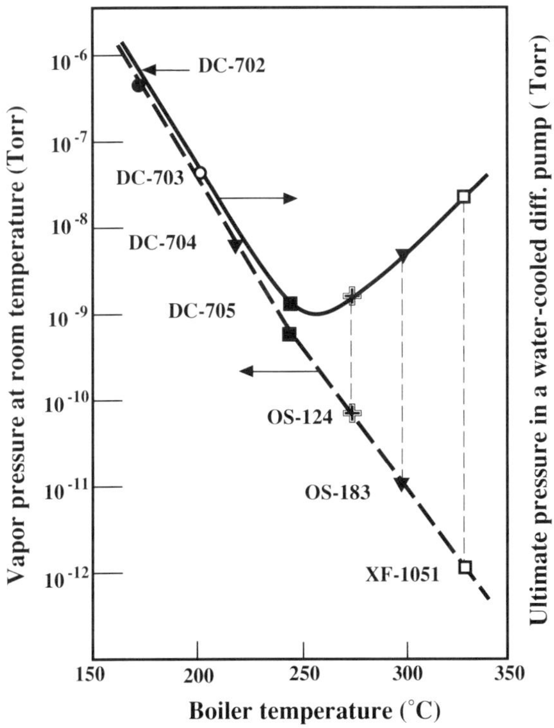


**Figure 9.8** Boiler temperature effect on the ultimate pressure of an diffusion pump using pumping fluids with successively lower vapor pressure. [After Hablanian (1974).]

tion, indicating that the maximum mass flow capacity of the pump has been reached. Finally, section 4 shows that all the pumping is accomplished only by the mechanical pump. The critical point on the curve assists in choosing the size of the backing (holding) pump (Hablanian, 1974).

*Diffusion pumps should not be used above the critical point.*

The boiler temperature effect on the ultimate pressure of water-cooled oil-diffusion pumps, using pumping fluids with successively lower pressures, is shown in Fig. 9.8.

For more details on diffusion pumps and other types of pumps, see Hablanian (1990).

## 9.4 Pumping by Dilution

The partial pressure of a gas-active component (e.g., $O_2$) in the gaseous environment (e.g., atmospheric air) within a certain volume can be reduced to a negligible amount, using the pumping-by-dilution technique. The technique consists of evacuating the vessel containing the active gas (usually at atmospheric pressure) to a low pressure, refilling the vessel with an inert gas to a certain pressure, and repeating this procedure several times. In the incandescent-lamp industry, where this technique is called "flushing," the diluting gas is nitrogen (selected for that purpose for economic reasons, although it is not inert). Thus

$$p_{pu} = p_1(p_f/p_r)^n \qquad \text{(e.g., mbar)} \tag{9.4}$$

where $p_{pu}$ (e.g., mbar) is the ultimate partial pressure of the chemically active gas obtained after the $n$th flushing; $p_1$ (e.g., mbar), the partial pressure of the chemically active gas after the first evacuation; $p_f$ (e.g., mbar), the pressure to which the vessel is evacuated repeatedly; $p_r$ (e.g., mbar), the pressure to which the vessel is filled with the neutral gas (or nitrogen); and $n$, the number of flushing cycles.

## Exercises

### Exercise 9.1

A rotary-vane vacuum pump, without gas ballast, evacuates a mixture of air and water vapor. The temperature of the pump is 55°C and the pressure $p_e$ required to raise the exhaust valve, quoted by the manufacturer of the pump, is 1400 mbar. What is the limit pressure above which the condensation of water vapor begins?

**Solution 9.1**

Use Eq. 9.2 written as

$$p_{v\,max}/p_g = p_{sv}/(p_e - p_{sv})$$

Substitute

$p_{sv} = 157.4$ mbar at 55°C (see Table 8.9)
$p_e = 1400$ mbar (quoted by the manufacturer of the pump)

Thus

$$p_{v\,max}/p_g = 155.4\text{ mbar}/(1400\text{ mbar} - 155.4\text{ mbar})$$
$$= 0.125$$

The condensation of water vapor will begin when

$$p_{v\,max} = 0.125p_g$$

i.e., when the partial pressure of water vapor at the port of the pump exceeds 12.5% of the air pressure.

**Exercise 9.2**

Air at atmospheric pressure is evacuated from an electric incandescent bulb down to a pressure of 1 mbar. Then the bulb is refilled with nitrogen at 100 mbar. This cycle is repeated 4 times. What is the amount of the residual oxygen in the gaseous environment of the bulb after the fourth cycle?

**Solution 9.2**

The partial pressure of oxygen in atmospheric air is $2.12 \times 10^2$ mbar (see Table 1.1). After evacuation of the bulb from atmospheric pressure to 1 mbar, the partial pressure of oxygen in the residual gas within the bulb is

$$p_1 = 2.12 \times 10^2\text{ mbar}/1013\text{mbar} = 0.209\text{ mbar}$$

Use Eq. 9.4:

$$p_{pu} = p_1(p_f/p_r)^n$$

Substitute

$p_1 = 0.209$ mbar
$p_f = 1$ mbar
$p_r = 100$ mbar
$n = 4$

Thus

$$p_u = 0.209 \text{ mbar } (1 \text{ mbar}/100 \text{ mbar})^4 = 2.09 \times 10^{-9} \text{ mbar}$$

## References

Hablanian, M. H. (1974), *Transactions of 6th International Vacuum Congress*, Kyoto, Japan.

Hablanian, M. H. (1979), in *Methods of Experimental Physics*, Vol. 14, *Vacuum Physics and Technology*, G. L. Weissler and R. W. Carlson (eds.), Academic Press, New York.

Hablanian, M. H. (1990), *High-Vacuum Technology*, Marcel Dekker, New York.

# 10

# Pumpdown Transient

## 10.1 *Constant-Speed Case, Gas Load Disregarded*

Conditions are as follows:

The pump and the vessel are directly connected (see Fig. 10.1).
The temperature of the vessel is constant.
The gas load, which sets the limit to the ultimate pressure of the system, is disregarded.

### 10.1.1 Evacuation Rate

$$Q = S_P p \qquad (\text{e.g., Pa m}^3\ \text{h}^{-1}) \tag{10.1}$$

$$= -V(dp/dt) \tag{10.1a}$$

where $Q$ is the flow rate of gas initially enclosed in the vessel being evacuated; $S_P$ (e.g., $m^3\ s^{-1}$), the pumping speed at the port of the pump, and hence at the port of the vessel; $p$ (e.g., Pa), the pressure of gas in the vessel; $V$ (e.g., $m^3$), the volume of the vessel; and $dp/dt$ (e.g., $m^3\ h^{-1}$), the rate of the pressure change with time in the vessel.

Equation 10.1 holds true if, and only if, $S_P$ is constant over the pressure range considered in calculations. Accurate results are obtained, e.g., for rotary-vane vacuum pumps whose pumping speed is fairly constant in the pressure range 1000–13 mbar (see Fig. 9.3).

### 10.1.2 Pressure Decay with Time

$$p_t = p_0 \exp[(-S_P/V)t] \qquad (\text{e.g., mbar}) \tag{10.2}$$

where $p_t$ (e.g., mbar) is the pressure of gas in the vessel after a time $t$ subsequent to the start of pumpdown; $p_0$ (e.g., mbar), the pressure of gas in the vessel at the start of pumpdown when $t = 0$; and $t$ (e.g., s) the time required to reach the pressure $p_t$.

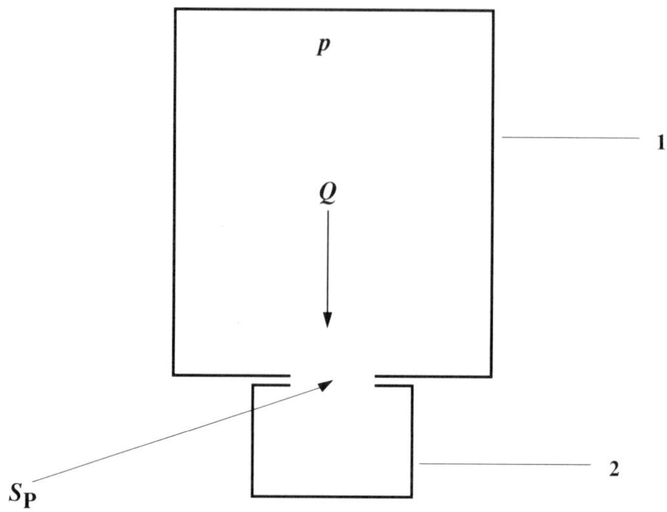


**Figure 10.1** Schematic of a vessel (1) and a pump (2) directly connected ($S_P$—pumping speed at the port of the pump, $Q$—flow rate of gas enclosed in the vessel, $p$—pressure of gas in the vessel).

### 10.1.3 Time Required to Reach a Certain Pressure

*a. Rotary vane, rotary piston, etc.*

$$t = (V/S_P)\ln(p_0/p_t) \qquad \text{(e.g., min)} \tag{10.3}$$

$$= 2.3(V/S_P)\log(p_0/p_t) \tag{10.3a}$$

The values of $t$ calculated by using Eq. 10.3 are accurate within a few seconds from the start of pumpdown until the pressure in the vessel reaches the value $p_t \cong 13$ mbar (10 Torr).

To obtain fairly accurate results for pressures $p_t < 13$ mbar, when outgassing becomes significant, multiply $t$ (Eq. 10.3) by

| | | |
|---|---|---|
| 1.5 | when | $6.7 \times 10^{-1}$ mbar $< p_t <$ 13 mbar |
| 2.0 | when | $6.7 \times 10^{-2}$ mbar $< p_t < 6.7 \times 10^{-1}$ mbar |
| 4.0 | when | $1.3 \times 10^{-3}$ mbar $< p_t < 6.7 \times 10^{-2}$ mbar |

*b. Roots pumps.*

$$t = (V/S_b)\ln[(p_0 + \Delta p)/(p_t + \Delta p)] \qquad \text{(e.g., min)} \qquad (10.4)$$

$$= 2.3(V/S_b)\log[(P_0 + \Delta p)/(p_t + \Delta p)] \qquad (10.4a)$$

where $S_b$ (e.g., liters min$^{-1}$) is the pumping speed of the backing pump; and $\Delta p$ (e.g., mbar), the differential pressure at the overflow valve (quoted by the manufacturer in the data sheet of the pump).

### 10.1.4 Time Required to Reduce Pressure to a Specified Value

*a. Time to reduce $p_0$ to $e^{-1}$ = 36.7% from its initial value.*

$$\tau = V/S_P \qquad \text{(e.g., s)} \qquad (10.5)$$

where $\tau$ is the time constant of the vessel.

*b. Half-life, or time to reduce $p_0$ to half its initial value.*

$$\tau_{1/2} = 0.693(V/S_P) \qquad \text{(e.g., s)} \qquad (10.6)$$

*c. Time to reduce $p_0$ by one decade.*

$$\tau_{1/10} = 2.3(V/S_P) \qquad \text{(e.g., s)} \qquad (10.7)$$

## *10.2 Constant-Speed Case, Gas Load Considered*

Conditions are as follows:

The pump and the vessel are directly connected (see Fig. 10.2).
The temperature of the vessel is constant.
A gas load, at almost constant rate, is considered.

### 10.2.1 Evacuation Rate

$$Q = S_P p - Q_G \qquad \text{(e.g., Torr s}^{-1}\text{)} \qquad (10.8)$$

where $Q$ is the flow rate of gas initially enclosed in the vessel; $S_P$ (e.g., liters s$^{-1}$), the pumping speed at the port of the pump, and hence at the port of the vessel; $p$ (e.g., Torr), the pressure of gas in the vessel; and $Q_G$ (e.g., Torr liters s$^{-1}$), the gas load at almost constant rate owing to outgassing, leakage, etc. (see Chapter 8).

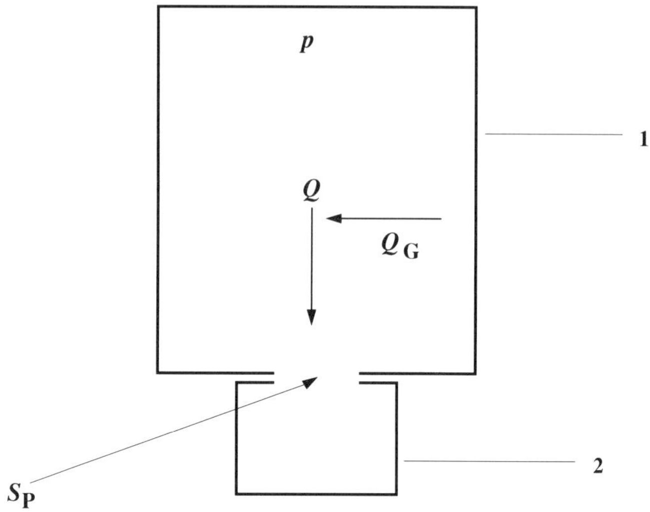


**Figure 10.2** Schematic of a vessel (1) and a pump (2) directly connected ($S_P$—pumping speed at the port of the pump, $Q$—flow rate of gas enclosed in the vessel, $p$—pressure of gas in the vessel, $Q_G$—gas load).

### 10.2.2 Pressure Decay with Time

$$p_t = (p_0 - p_u)\exp[(S_P/V) + p_u] \quad \text{(e.g., mbar)} \qquad (10.9)$$

where $p_t$ (e.g., mbar), is the gas pressure in the vessl subsequent to the time $t$ from the start of pumpdown; $p_0$ (e.g., Torr), the gas pressure at the start of pumpdown, when $t = 0$; $p_u$ (e.g., Torr), the ultimate pressure of the system due to the gas load; $V$ (e.g., liters), the volume of the vessel; and $t$ (e.g., s), the time required to reach the pressure $p_t$.

### 10.2.3 Time Required to Reach a Certain Pressure

$$t = (V/S_P)\ln[(p_0 - p_u)/(p_t - p_u)] \quad \text{(e.g., s)} \qquad (10.10)$$

$$= 2.3(V/S_P)\log[(p_0 - p_u)/(p_t - p_u)] \qquad (10.10a)$$

where the quantities have the same meaning as in Eq. 10.9.

The nomogram illustrated in Fig. 10.3 permits the evaluation of time $t$ required to reduce the air pressure, at room temperature, in a vessel, from 10

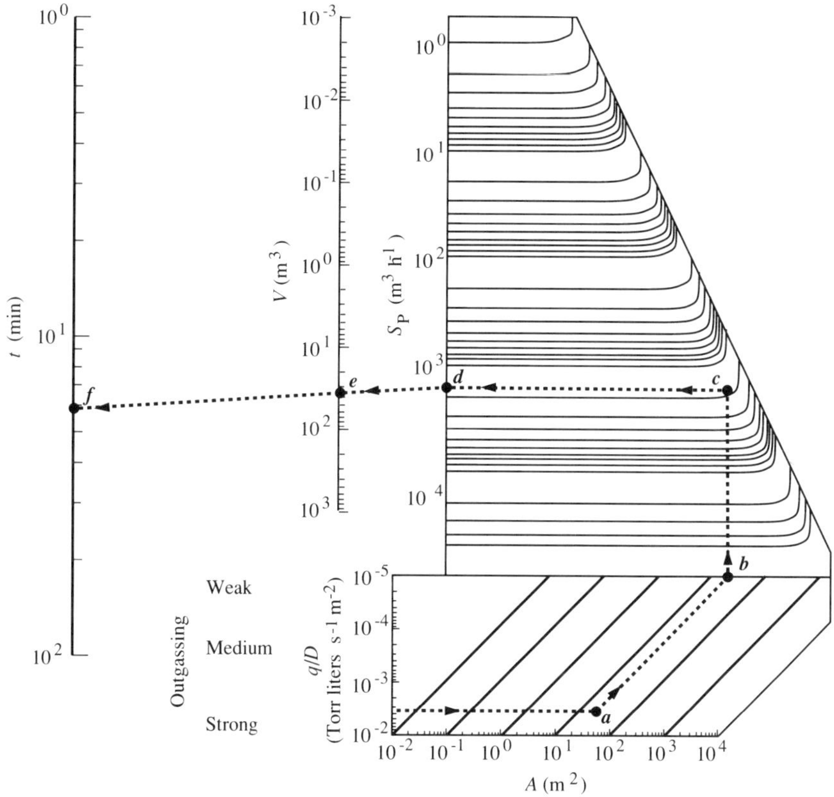


**Figure 10.3** Nomogram for the evaluation of time required to reduce the air pressure, at room temperature, from 10 Torr to $10^{-3}$ Torr. The gas load contributed by the walls of the vessel is considered. The pump and the vessel are directly connected. [After Diels and Jaeckel (1958).]

Torr to $10^{-3}$ Torr, the gas load produced by the outgassing of the walls being considered.

## 10.3 *Variable-Speed Case, Gas Load Disregarded*

Conditions are as follows:

The pump and the vessel are directly connected.
The temperature of the vessel is constant.
The gas load, which sets the limit to the ultimate pressure of the system, is disregarded.

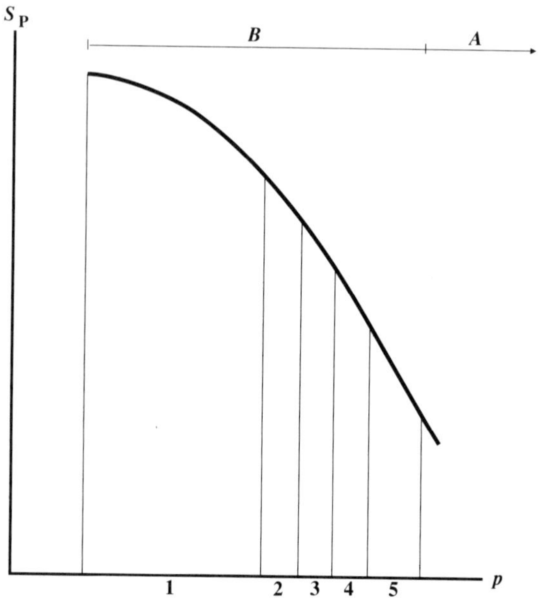


**Figure 10.4** Performance characteristic of a Roots pump: A—region of constant speed (backing pump), B—region of variable pumping speed.

### 10.3.1 Time Required to Reduce Pressure to a Certain Value

Determine intervals, on the performance characteristic of the pump, characterized by approximately constant pumping speed (see Fig. 10.4).

Use Eq. 10.3a to calculate the pumpdown time for each interval.

Add up the pumping speeds obtained for each interval. Thus

$$t_5 = 2.3(V/S_{P\,5-4})\log(p_5/p_4) \qquad \text{(e.g., s)} \tag{10.11}$$

$$t_4 = 2.3(V/S_{P\,4-3})\log(p_4/p_3) \tag{10.11a}$$

$$\vdots$$

$$t_n = \sum_{1}^{n}(t_{1-n}) \tag{10.11b}$$

## 10.4 Constant-Throughput Case

### 10.4.1 Time Required to Reach a Certain Pressure

The time during which diffusion pumps work within the constant-throughput section of the performance characteristic (see Fig. 9.7) is

$$t = (V/Q)(p_0 - p_t) \qquad \text{(e.g., s)} \tag{10.12}$$

This time is very short, less than 10 s for a typical bell-jar system (Hablanian, 1979, p. 106).

## 10.5 Pumpdown Transient in the Viscous Range

Conditions are as follows:

The pump and the vessel are connected by vacuum components (pipe, valve, baffle, etc.).
The speed of the pump is constant in the range of pressures considered.

### 10.5.1 Time Required to Reach a Certain Pressure in the Viscous Range

$$\begin{aligned} t = {} & V/\zeta(1/p_t - 1/p_0) \\ & + V/S_P\left\{\left[(S_P/\zeta)^2 + p_t^2\right]^{1/2}\Big/p_t - \left[(S_P/\zeta)^2 + p_0^2\right]^{1/2}\Big/p_0\right\} \\ & + V/S_P\left(\ln\left\{p_0 + \left[(S_P/\zeta)^2 + p_0^2\right]^{1/2}\right\}\right)\Big/ \\ & \left\{p_t + \left[(S_P/\zeta)^2 + p_t^2\right]^{1/2}\right\} \qquad \text{(s)} \end{aligned} \tag{10.13}$$

with $\zeta = (\zeta/128)(d^4/\eta L)$, $\eta$ (poise), $d$ (cm), $L$ (cm) the diameter and length of the pipe connecting the pump to the vessel, $p_0$ (dyn cm$^{-2}$), $p_t$ (dyn cm$^{-2}$), and $S_P$ (cm$^3$ s$^{-1}$).

Equation 10.130 holds true if the volume of the pipe, which connects the pump to the vessel is negligible compared to that of the vessel, and the flow rates of gas leaving the vessel and entering the pump are equal at each instant.

The graph in Fig. 10.5, based on Eq. 10.13, permits the evaluation of the time required to decrease the pressure from 760 Torr to $7.6 \times 10^{-2}$ Torr in a volume $V$.

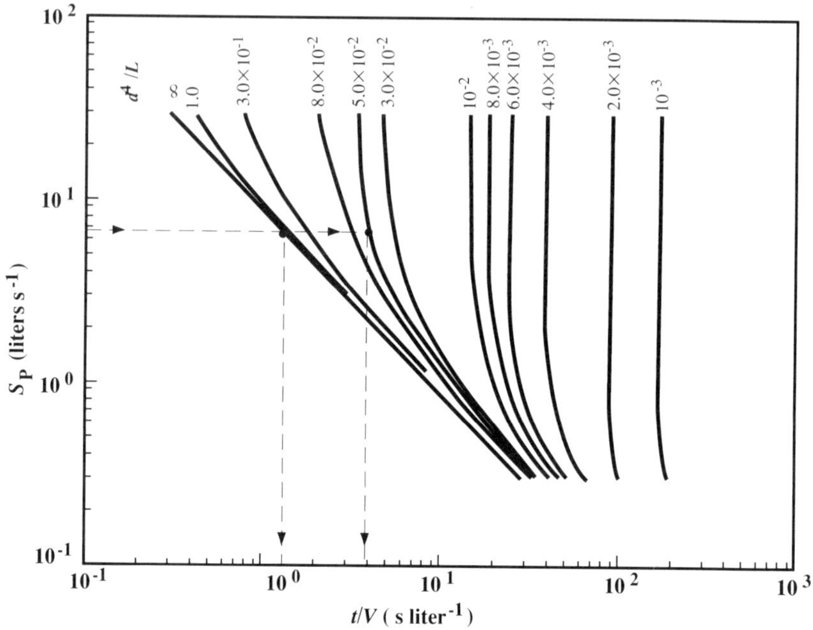


**Figure 10.5** Time required to reduce the air pressure in a vessel at room temperature from 760 Torr to $7.6 \times 10^{-2}$ Torr. The vessel of volume $V$ is connected to the pump by a pipe of diameter $d$ (cm) and length $L$ (cm). The pumping speed is $S_P$. [After Delafosse and Mongodin (1961).]

## *10.6 Pumpdown Transient in the Molecular Range*

Conditions are as follows:

The pump and the vessel are connected by vacuum components (pipe, valve, baffle, etc.; see Fig. 10.6).
The speed of the pump is constant with pressure.
A gas load at almost constant rate is considered.

### 10.6.1 Time Required to Reach a Certain Pressure in the Molecular Range

$$t = Q_{Dm}t'/[S_n(p_t - p_u - Q_W/S_n)] + Q_{D\Sigma}t'/[S_n(p_t - p_u - Q_W/S_n)]^2 \quad \text{(s)} \qquad (10.14)$$

where $Q_{Dm}$ (e.g., mbar liter s$^{-1}$) is the gas flow at almost constant rate owing to the outgassing of metallic surfaces exposed to vacuum at time $t'$; $t'$ (e.g., s),

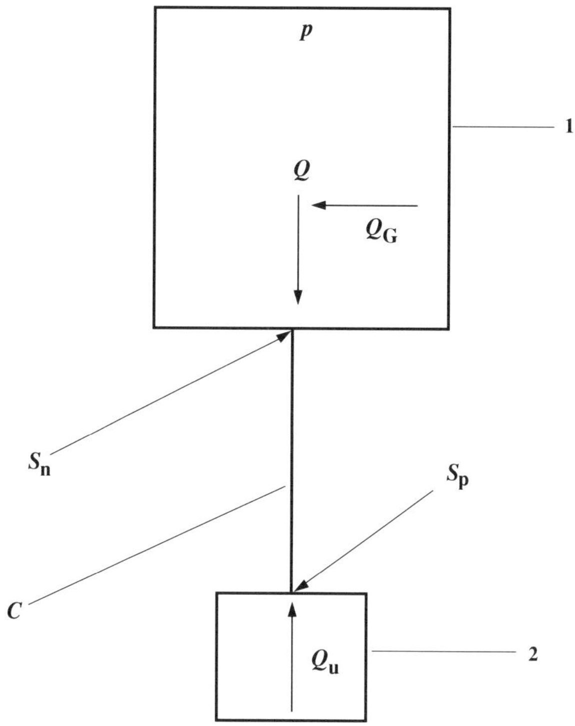


**Figure 10.6** Schematic of a vessel (1) and a pump (2) connected by a pipe of conductance C [$S_P$—pumping speed at the port of the pump, $S_n$—net (effective) pumping speed at the path of the vessel, $Q_u$—flow rate of gas returned from the pump to the vessel, Q—flow rate of gas enclosed in the vessel, $Q_G$—gas load, p—pressure in the vessel].

outgassing time, usually 3600 s; $S_n$ (e.g., liters $s^{-1}$), the pumping speed at the port of the vessel; $p_t$ (e.g., mbar), the pressure of gas in the vessel at the start of pumpdown when $t = 0$; $p_u$ (e.g., mbar), the ultimate pressure of the pump $Q_W$ (e.g., mbar liters $s^{-1}$), the gas flow at almost constant rate owing to leaks, permeation, and residual gases set free during the process in vacuum; and $Q_{D\Sigma}$ (e.g., mbar liters $s^{-1}$), the gas flow at almost constant rate owing to the outgassing of elastomers, polymers, and glass.

The gas flow rates $Q_{Dm}$, $Q_W$, and $Q_{D\Sigma}$ are calculated as

$$Q_{Dm} = q_{Dm} A \qquad (\text{e.g., mbar liters s}^{-1}) \tag{10.15}$$

where $q_{Dm}$ (e.g., mbar $s^{-1}$ $cm^{-2}$) is the specific outgassing rate of metallic surfaces (see Section 8.2.1); and $A$ (e.g., $cm^2$) the total outgassing area. Thus

$$Q_W = Q_L + Q_{Per} + Q_r \qquad \text{(e.g., mbar liters s}^{-1}\text{)} \tag{10.16}$$

where $Q_L$ (e.g., mbar liters $s^{-1}$), $Q_{Per}$ (e.g., mbar liters $s^{-1}$), $Q_r$ (e.g., mbar liters $s^{-1}$) are the flow rates due to leaks (see Section 8.2.3), permeation (see Section 8.2.4), and residual gases set free from the process in vacuum, such as evaporation (see Section 8.2.2). Thus

$$Q_D = (q_{De} + q_{poly} + q_{glass})A \qquad \text{(e.g., mbar liters s}^{-1}\text{)} \tag{10.17}$$

where $q_{De}$ (e.g., mbar liters $s^{-1}$ $cm^{-2}$) is the specific outgassing rate of elastomers (see Table 8.2); $g_{poly}$ (e.g., mbar liters $s^{-1}$ $cm^{-2}$), the specific outgassing rate of polymers (see Table 8.3); and $g_{glass}$ (e.g., mbar liters $s^{-1}$ $cm^{-2}$), the specific outgassing rate of glass and ceramics (see Table 8.4).

## *Exercises*

**Exercise 10.1**

Work out a practical formula to calculate the time requested to a pump of constant pumping speed $S_P$ to evacuate a vessel from atmospheric pressure down to 13 mbar. Assume that the pump and the vessel of volume $V$ are directly connected and the temperature of the vessel is maintained constant during the pumpdown process.

**Solution 10.1**

Use Eq. 10.3a

$$t = 2.3(V/S_P)\log(p_0/p_t)$$

and substitute

$$p_0 = 1 \text{ atm} = 1013 \text{ mbar}$$

$$p_t = 13 \text{ mbar}$$

Thus

$$t = 2.3(V/S_P)\log(1013 \text{ mbar}/13 \text{ mbar}) = 4.35(V/S_P) \qquad \text{(s)}$$

This rule of thumb is valid for pumps whose pumping speed is constant in the range of pressures 1013–13 mbar (see, e.g., the performance characteristics of double-stage rotary-vane vacuum pumps in Fig. 9.3).

**Exercise 10.2**

Calculate the time required to reach $1/n$ of the initial pressure of gas in a vessel of volume $V$, evacuated by a pump of constant pumping speed $S_P$. The pump and the vessel are directly connected and the temperature of the vessel maintained constant. Neglect the gas load.

**Solution 10.2**

Use Eq. 10.3a

$$t_{1/2} = 2.3(V/S_P)\log(p_0/p_t) \qquad (\text{s})$$

For $p_t = p_0/2$:

$$t_{1/2} = 2.3(V/S_P)\log[p_0/(p_0/2)]$$

$$= 2.3(V/S_P)\log 2 = 0.693(V/S_P) \qquad (\text{s}) \qquad (\text{see Eq. 10.6})$$

For $p_t = p_0/2^2$:

$$t_{1/4} = 2.3(V/S_P)\log[p_0/(p_0/4)]$$

$$= 2.3(V/S_P)\log 4 = 1.385(V/S_P) \qquad (\text{s})$$

For $p_t = p_0/2^n$:

$$t_{1/n} = 2.3(V/S_P)\log[p_0/(p_0/2^n)]$$

$$= 2.3(V/S_P)n\log 2 = 0.693N(V/S_P) \qquad (\text{s})$$

Generally

$$t_{1/n} = nt_{1/2} = 0.693n(V/S_P) \qquad (\text{s})$$

holds true as long as $S_P = \text{const}$. Fairly accurate results are obtained for rotary-vane vacuum pumps in the pressure range 1013–13 mbar. Since

$$p_t = p_0/2^n$$

there results

$$n = \log(p_0/p_t)/\log 2$$

On substituting $p_0 = 1013$ mbar and $p_t = 13$ mbar

$$n = \log(1013\ \text{mbar}/13\ \text{mbar})/\log 2 = 6.28$$

Thus

$$t_{1/n} = 0.693n(V/S_P) = 0.693 \times 6.28(V/S_P) = 4.35(V/S_P)$$

See also Exercise 10.1.

**Exercise 10.3**

A rotary-vane vacuum pump connected directly to a 100-liter vessel evacuates it from atmospheric pressure to 10 mbar. Then the pump is valved off and the vessel left for 5 min 30 s to process the parts inside. At the end of this interval, the pressure in the vessel is raised in 1 min to atmospheric pressure, and left open for 5 minutes. Soon after, the vessel is shut off and the cycle described above is repeated. Calculate the size of the pump required to meet these conditions, assuming that the vessel must be recycled 5 times per hour.

**Solution 10.3**

For completing a cycle it is necessary to

Pump down the vessel from 1013 mbar to 10 mbar ($x$ min)
Accomplish the process in the vessel at 10 mbar (5 min 30 s)
Raise the pressure in the vessel from 10 to 1013 mbar (1 min)
Open the vessel to the atmosphere for (5 min)
Total time per cycle ($x$ + 11 min 30 s)
Time to accomplish a cycle (60 min/5 cycles = 12 min)

The time left for evacuating the vessel from 1013 mbar to 10 mbar is therefore

$$12 \text{ min} - (x + 11 \text{ min } 30 \text{ s})$$

Thus $x = 30$ s. Use Eq. (10.3a) written as

$$S_P = 2.3(V/t)\log(p_0/p_t)$$

Substitute

$V = 100$
$t = 30$ s
$p_0 = 1013$ mbar
$p_t = 10$ mbar

Thus

$$S_P = 2.3(100/30)\log(1013/10) = 15.4 \text{ liters s}^{-1}$$

$$= 15.4 \text{ liters s}^{-1} \times 10^{-3} \text{ m}^3 \text{ h}^{-1} \times 3600 \text{ s h}^{-1} = 55.4 \text{ m}^3 \text{ h}^{-1}$$

Theoretically, it would be sufficient to choose a rotary-vane vacuum pump this size. In practice, however, one selects the next size of pump in the row of standard pumps, say, 100 $m^3$ $h^{-1}$.

**Exercise 10.4**

A vessel of 6 $m^3$ containing air at atmospheric pressure must be evacuated to 10 mbar in 10 min by a pump a given distance away. Select the pump and its size, assuming that the temperature of the vessel is constant during the pumpdown process.

**Solution 10.4**

(i) Assume that the pump is connected to the port of the vessel. Therefore Eq. 10.3a applies, written as

$$S_P = 2.3(V/t)\log(p_0/p_t)$$

Substitute

$V = 6\ \text{m}^3$
$t = 10\ \text{min} = 0.6\ \text{h}$
$p_0 = 1013\ \text{mbar}$
$p_t = 10\ \text{mbar}$

Hence

$$S_p = 2.3(6/0.6)\log(1013/10) = 46\ \text{m}^3\ \text{h}^{-1}$$

(ii) For the efficient utilization of pump capacity, the pressure drop along the pipe that connects the pump to the vessel must not exceed 20% of the pressure at the port of the pump, at operating level (see Section 5.3.3). Thus, if $S_n$ is the net pumping speed at the port of the vessel

$$S_n/S_P = 1 - 0.2$$

$$S_P = S_n/0.8 = 46\ \text{m}^3\ \text{h}^{-1}/0.8 = 57.5\ \text{m}^3\ \text{h}^{-1}$$

According to Fig. 9.2, the mechanical pumps are the most economical choice for pumping down from atmospheric pressure to 10 mbar. One selects, therefore, a rotary-vane vacuum pump in excess of 57.5 $\text{m}^3\ \text{h}^{-1}$, say, 75 $\text{m}^3\ \text{h}^{-1}$.

**Exercise 10.5**

A Roots pump backed by a double-stage rotary-vane vacuum pump of 75 $\text{m}^3\ \text{h}^{-1}$ is directly connected to a vessel of 100 $\text{m}^3$. The differential pressure at the overflow valve of the Roots pump is 40 mbar. Use the performance characteristic curve of the pump supplied by the manufacturer to calculate the time required to evacuate the vessel from atmospheric pressure down to $10^{-2}$ mbar.

**Solution 10.5** Refer to Fig. 10A.

(i) The performance characteristics curve $S_P = f(p)$ of roots pumps shows two distinct fields: A, where the backing pump accomplishes the evacuation of the vessel; and B, where the Roots pump takes over the pumpdown process. The time required to evacuate the vessel from atmospheric pressure to $10^{-2}$ mbar is therefore the sum of times to pump fields A and B.

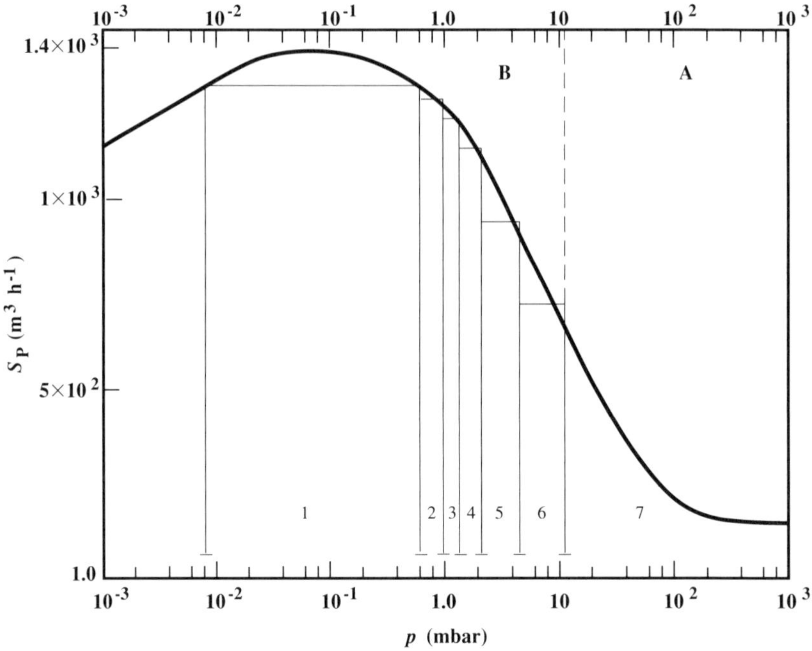


Figure 10A

(ii) To determine the time required by the backing pump to pump down from $10^3$ mbar to 10 mbar in field A, use Eq. 10.3a written as

$$t_A = 2.3(V/S_b)\log\left[(p_0 + \Delta p)/(p_t + \Delta p)\right]$$

Substitute

$V = 100\ \mathrm{m}^3$
$S_b = 75\ \mathrm{m}^3\ \mathrm{h}^{-1}$
$p_0 = 10^3$ mbar
$p_t = 10$ mbar
$\Delta p = 40$ mbar

Hence

$$t_A = 2.3(100/75)\log[(1000 + 40)/(10 + 40)] = 4.04\ \mathrm{h}$$

(ii) Estimate the time required by the Roots pump to complete the evacuation of the vessel from 10 mbar to $10^{-2}$ mbar in field B. Divide the field B of the performance characteristic curve in several ranges of pressure within which

$S_P$ changes just a little, and calculate the time requested to pump down in each range, using Eq. 10.3a. Add up the pumping times calculated for each range to estimate the total time required to evacuate the vessel in field B. Substitute in Eq. 10.3a, for each region, the following data:

*Region 6*

$V = 100\ \mathrm{m}^3$
$S_P = 7.2 \times 10^2\ \mathrm{m}^3\ \mathrm{h}^{-1}$ (read on the performance characteristic curve)
$p_0 = 10$ mbar
$p_t = 4$ mbar

Hence

$$t_6 = 2.3(100/7.2 \times 10^2)\log(10/4) = 1.27 \times 10^{-1}\ \mathrm{h}$$

*Region 5*

$V = 100\ \mathrm{m}^2$
$S_P = 9.4 \times 10^{-2}\ \mathrm{m}^3\ \mathrm{h}^{-1}$ (read on the performance characteristic curve)
$p_0 = 4.0$ mbar
$p_t = 2.0$ mbar

Hence

$$t_5 = 2.3(100/9.4 \times 10^{-2})\log(4/2) = 7.37 \times 10^{-2}\ \mathrm{h}$$

*Region 4*

$V = 100\ \mathrm{m}^3$
$S_P = 1.13 \times 10^3\ \mathrm{m}^3\ \mathrm{h}^{-1}$ (read on the performance characteristic curve)
$p_0 = 2.0$ mbar
$p_t = 1.3$ mbar

Hence

$$t_4 = 2.3(100/1.13 \times 10^3)\log(2/1.3) = 3.80 \times 10^{-2}\ \mathrm{h}$$

*Region 3*

$V = 100\ \mathrm{m}^3$
$S_P = 1.21 \times 10^3\ \mathrm{m}^3\ \mathrm{h}^{-1}$ (read on the performance characteristic curve)
$p_0 = 1.3$ mbar
$p_t = 9.3 \times 10^{-1}$ mbar

Hence

$$t_3 = 2.3(100/1.21 \times 10^3)\log(1.3/9.3 \times 10^{-1}) = 2.76 \times 10^{-2}\ \mathrm{h}$$

*Region 2*

$V = 100\ \text{m}^3$
$S_P = 1.26 \times 10^3\ \text{m}^3\text{h}^{-1}$ (read on the performance characteristic curve)
$p_0 = 9.3 \times 10^{-1}$ mbar
$p_t = 5.6 \times 10^{-1}$ mbar

Hence

$$t_2 = 2.3(100/1.26 \times 10^3)\log(9.3 \times 10^{-1}/5.6 \times 10^{-1})$$

$$= 4.02 \times 10^{-2}\ \text{h}$$

*Region 1*

$V = 100\ \text{m}^3$
$S_P = 1.30 \times 10^3$ mbar (read on the performance characteristic curve)
$p_0 = 5.6 \times 10^{-1}$ mbar
$p_t = 10^{-2}$ mbar

Hence

$$t_1 = 2.3(100/1.3 \times 10^3)\log(5.6 \times 10^{-1}/10^{-2}) = 3.09 \times 10^{-1}\ \text{h}$$

Thus, time to pump down field B is

$$t_B = \sum_1^6 t_{1-6} = 3.09 \times 10^{-1}\ \text{h} + 4.02 \times 10^{-2}\ \text{h} + 2.76 \times 10^{-2}\ \text{h}$$

$$+ 3.80 \times 10^{-2}\ \text{h} + 7.37 \times 10^{-2}\ \text{h} + 1.27 \times 10^{-1}\ \text{h}$$

$$= 0.6155\ \text{h}$$

and time to pump the vessel from atmosphere down to $10^{-2}$ mbar is

$$t = t_A + t_B = 4.04\ \text{h} + 0.6155\ \text{h} = 4.6555\ \text{h}$$

**Exercise 10.6**

A 250-m$^3$ h$^{-1}$ double-stage rotary-vane vacuum pump evacuates a vessel in which atmospheric air leaks at a rate of $6.11 \times 10^{-2}$ mbar liters s$^{-1}$. What is the reduction of the pumping speed of the pump caused by that leak at 1, $10^{-1}$,

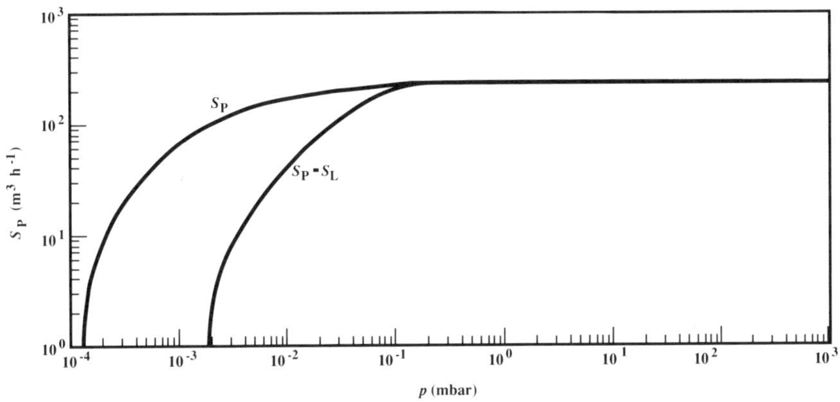


Figure 10B

$10^{-2}$, $4 \times 10^{-3}$, and $2 \times 10^{-3}$ mbar? Assume that the pump is directly connected to the vessel.

**Solution 10.6**

Refer to Fig. 10B.

(i) The speeds $S_P$ at the port of the pump, corresponding to the pressure requested, read on the performance characteristic curve, are as follows:

| $p$ (mbar) | $S_P$ ($m^3\ h^{-1}$) |
|---|---|
| 1.0 | 250 |
| $1.0 \times 10^{-1}$ | 220 |
| $1.0 \times 10^{-2}$ | 180 |
| $4.0 \times 10^{-3}$ | 150 |
| $2.0 \times 10^{-3}$ | 110 |

(ii) The flow rate of atmospheric air into the vessel, due to the leak, is

$$Q_L = 6.11 \times 10^{-2} \text{ mbar liters s}^{-1} \times 10^{-3} \text{ m}^3 \text{ h}^{-1} \times 3600 \text{ s h}^{-1}$$

$$= 0.22 \text{ mbar m}^3 \text{ h}^{-1}$$

and the volume rate of flow due to the leakage (see Eq. 4.38a) is

$$S_L = Q_L/p$$

Thus for

| | |
|---|---|
| 1.0 mbar | $S_L = 0.22\ \text{mbar m}^3\ \text{h}^{-1}/1.0\ \text{mbar} = 0.22\ \text{m}^3\ \text{h}^{-1}$ |
| $1.0 \times 10^{-1}$ mbar | $S_L = 0.22\ \text{mbar m}^3\ \text{h}^{-1}/1.0 \times 10^{-1}\ \text{mbar} \cong 2.2\ \text{m}^3\ \text{h}^{-1}$ |
| $1.0 \times 10^{-2}$ mbar | $S_L = 0.22\ \text{mbar m}^3\ \text{h}^{-1}/1.0 \times 10^{-2}\ \text{mbar} = 22\ \text{m}^3\ \text{h}^{-1}$ |
| $4.0 \times 10^{-3}$ mbar | $S_L = 0.22\ \text{mbar m}^3\ \text{h}^{-1}/4.0 \times 10^{-3}\ \text{mbar} = 55\ \text{m}^3\ \text{h}^{-1}$ |
| $2.0 \times 10^{-3}$ mbar | $S_L = 0.22\ \text{mbar m}^3\ \text{h}^{-1}/2.0 \times 10^{-3}\ \text{mbar} = 110\ \text{m}^3\ \text{h}^{-1}$ |

(iii) The reduction of the pumping speed $S_P - S_L$ is

| | |
|---|---|
| 1.0 mbar | $250\ \text{m}^3\ \text{h}^{-1} - 0.22\ \text{m}^3\ \text{h}^{-1}$ = negligible, small |
| $1.0 \times 10^{-1}$ mbar | $220\ \text{m}^3\ \text{h}^{-1} - 2.2\ \text{m}^3\ \text{h}^{-1} = 218\ \text{m}^3\ \text{h}^{-1}$ |
| $1.0 \times 10^{-2}$ mbar | $180\ \text{m}^3\ \text{h}^{-1} - 22\ \text{m}^3\ \text{h}^{-1} = 158\ \text{m}^3\ \text{h}^{-1}$ |
| $4.0 \times 10^{-3}$ mbar | $150\ \text{m}^3\ \text{h}^{-1} - 55\ \text{m}^3\ \text{h}^{-1} = 95\ \text{m}^3\ \text{h}^{-1}$ |
| $2.0 \times 10^{-3}$ mbar | $110\ \text{m}^3\ \text{h}^{-1} - 150\ \text{m}^3\ \text{h}^{-1} = 0\ \text{m}^3\ \text{h}^{-1}$ |

**Exercise 10.7**

A 18-$\text{m}^3\ \text{h}^{-1}$ double-stage rotary-vane vacuum pump evacuates a 75 bell-jar from atmospheric pressure down to 10 mbar. The bell-jar, whose temperature is maintained constant at $T = 23°\text{C}$, contains 6 g of water. Calculate the pumpdown time, given that the pump is directly connected to the bell-jar.

**Solution 10.7**

The time required to pump down the bell-jar from atmospheric pressure to 10 mbar is

$$t = t_1 + t_2$$

where $t_1$ is the time required to evacuate the bell-jar disregarding the amount of water vapor inside it, and $t_2$ is the time necessary to evacuate the water vapor.

(i) Calculation of $t_1$. Use Eq. 10.3a written as

$$t_1 = 2.3(V/S_P)\log(p_0/p_t)$$

Substitute

$V = 75$ liters
$S_P = 18\ \text{m}^3\ \text{h}^{-1} \times 10^3$ liters $\text{m}^{-3} \times (1/3600)\ \text{h s}^{-1} = 5$ liters $\text{s}^{-1}$
$p_0 = 1013$ mbar
$p_t = 10$ mbar

Hence

$$t_1 = 2.3(75/5)\log(1013/10) = 69.2\ \text{s}$$

(ii) Calculation of $t_2$. The pressure in the bell-jar is continuously decreased by the pump until the pressure of the system is determined by the evaporation pressure of the water. When evaporation begins, the water vapor and liquid water are in equilibrium; i.e., the pressure remains constant until the evaporation surface of the water is reduced to the last molecular layer. Then, the pressure will fall as predicted by Eq. 10.2. The pressure $p_{sv}$ of saturated water vapor at $T = 23°\text{C}$ (see Table 8.8) is

$$p_{sv} = 28.088\ \text{mbar}$$

and, from Eq. 1.12a

$$V_{H_2O} = (W_t/M)RT/p_{sv}$$

Substitute

$W_t = 6$ g
$M = 18\ \text{g mol}^{-1}$ (see Table 1.2)
$R = 83.14$ mbar liters $\text{mol}^{-1}\ \text{K}^{-1}$ (see Table 1.3)
$T = 273 + 23 = 296$ K
$p_{sv} = 28.088$ mbar

Thus the volume of water vapor is

$$V_{H_2O} = (6\ \text{g}/18\ \text{g mol}^{-1}) \times 83.14\ \text{mbar liters mol}^{-1}\ \text{K}^{-1}$$
$$\times 296\ \text{K}/28.088\ \text{mbar} = 292\ \text{liters}$$

and

$$t_2 = V_{H_2O}/S_P = 292\ \text{liters}/5\ \text{liters s}^{-1} = 58.4\ \text{s}$$

The total time to evacuate the vessel is

$$t = t_1 + t_2 = 69.2\ \text{s} + 58.4\ \text{s} = 127.6\ \text{s}$$

**Exercise 10.8**

A 50-$m^3$ tank is directly connected to a 1750-$m^3$ $h^{-1}$ vacuum pump. The surface area of the tank walls is 85 $m^2$, and the specific outgassing rate of the material of the wall is $3.6 \times 10^{-3}$ Torr liters $s^{-1}$ $m^{-2}$. Use both the nomogram in Fig. 10.3 and the analytical calculation to determine the time required to evacuate the tank from 10 Torr down to $10^{-3}$ Torr.

**Solution 10.8**

(i) Evaluation of the pumpdown time according to the nomogram in Fig. 10.3 (interrupted line). The lines passing through $3.6 \times 10^{-3}$ Torr liters $s^{-1}$ $m^{-2}$ (outgassing scale $q_D$) and 85 $m^2$ (wall area, scale A) determine point *a*. Join point *a* to point *b* by an upward-sloping line and then raise a vertical line to point *c* upward to the curve corresponding to a pumping speed of $1.5 \times 10^3$ $m^3$ $h^{-1}$ (scale $S_P$). If point *c* falls inside the border curve area, the evaluated pumping speed $S_P$ can handle the outgassing rate selected on the scale $q_D$. The line joining point *d* on the scale $S_P$ with 50 $m^3$ (the volume of the tank) on scale *V* intersects scale *t* (pumpdown time) at point $f = 17$ min. Thus, the time required to pump the tank from 10 Torr to $10^{-3}$ Torr, in the presence of the outgassing from the walls, is 17 min.

(ii) Calculation of the pumping time. Use Eqs. 4.38a, 8.2, and 10.10a written as

$$p_D = Q_D/S_P$$

$$Q_D = q_D A$$

$$t = 2.3(V/S_P)\log[(p_0 - p_D)/(p_t - p_D)]$$

Substitute

$q_D = 3.6 \times 10^{-3}$ Torr
$A = 85$ $m^2$
$V = 50$ $m^3 = 50 \times 10^3$ liters
$S_P = 1750$ $m^3$ $h^{-1} \times 10^{-3}$ liters $m^{-3} \times (1/3600)$ h $s^{-1} = 486$ liters $s^{-1}$
$p_0 = 10$ Torr
$p_t = 10^{-3}$ Torr

Calculate

$$Q_D = 3.6 \times 10^{-3} \text{ Torr liters s}^{-1} \text{ m}^{-2} \times 85 \text{ m}^2 = 0.306 \text{ Torr liters s}^{-1}$$

$$p_D = 0.306 \text{ Torr liters s}^{-1}/486 \text{ liters s}^{-1} = 6.3 \times 10^{-4} \text{ Torr}$$

Hence

$$t = 2.3(50 \times 10^3/486)\log[(10 - 6.3 \times 10^4)/(1 \times 10^{-3} - 6.3 \times 10^{-4})]$$
$$= 1049\ \text{s} = 17.5\ \text{min}$$

**Exercise 10.9**

Use Eq. 10.13 to calculate the time required for a positive-displacement pump directly connected to a vessel to evacuate the vessel in the viscous range of flow.

**Solution 10.9**

From Eq. 10.13

$$t = V/\zeta(1 - p_t - 1/p_0)$$
$$+ V/S_P\left\{\left[(S_P/\zeta)^2 + p_t^2\right]^{1/2}\Big/p_t - \left[(S_P/\zeta)^2 + p_0^2\right]^{1/2}\Big/p_0\right\}$$
$$+V/S_P \ln\left\{p_0 + \left[(S_P/\zeta)^2 + p_0^2\right]^{1/2}\right\}\Big/\left\{p_t + \left[(S_P/\zeta)^2 + p_t^2\right]^{1/2}\right\}$$

where

$$\zeta = (\pi/128)(d^4/\eta L)$$

Since the pump and the vessel are directly connected, $L = 0$ and $\zeta = \infty$. Thus

$$t = (V/S_P)\ln(P_0/p_t)$$

which is Eq. 10.3.

**Exercise 10.10**

A 24-$m^3\ h^{-1}$ positive-displacement vacuum pump evacuates a 75-liter vessel maintained at $T = 25°C$. Use the graph in Fig. 10.5 to determine the time required to pump the air at atmospheric pressure from the vessel down to $7.5 \times 10^{-2}$ Torr. Consider the following possibilities: (a) the vessel is connected to the pump by a pipe 2.10 m long and of 1.8 cm inner diameter, and (b) the pump is connected directly to the vessel.

**Solution 10.10**

(a) Calculate the parameter $d^4/L$ of the diagram using $d = 1.8$ cm, $L = 210$ cm. Thus

$$d^4/L = 1.8^4/210 = 5 \times 10^{-2}$$

The pumping speed of the pump is

$$S_P = 24\ \text{m}^3\ \text{h}^{-1} \times 10^3\ \text{liters m}^{-3} \times (1/3600)\ \text{h s}^{-1} = 6.7\ \text{liters s}^{-1}$$

The horizontal (broken) line through $S_P = 6.7$ liters s$^{-1}$ intersects the curve $d^4/L = 5.9 \times 10^{-2}$ at 3.9 s liter$^{-1}$ measured on the scale $t/V$. Hence

$$t/V = 3.9 \text{ s liter}^{-1}$$

and with $V = 75$ liters

$$t = 3.9 \text{ s liter}^{-1} \times 75 \text{ liters} = 292.5 \text{ s}$$

Taking into account the delay of the air flow from the vessel to the pump, owing to the conductance of the pipe, as well as the inaccuracies of the readings, consider

$$t = 315 \text{ s}$$

(b) Since $L = 0$, $d^4/L = \infty$. The horizontal (broken) line through $S_P = 6.7$ s$^{-1}$ intersects the curve $d^4/L = \infty$ at 1.2 s liters$^{-1}$, measured on the scale $t/V$. Hence

$$t/V = 1.2 \text{ s liter}^{-1}$$

and with $v = 75$ liters

$$t = 1.2 \text{ s liter}^{-1} \times 75 \text{ liters} = 90 \text{ s}$$

Taking into account the inaccuracies of the graph readings, consider

$$t = 100 \text{ s}$$

**Exercise 10.11**

Calculate the time required for an oil-diffusion pump to evacuate air from a stainless-steel vessel in the range $4 \times 10^{-3}$ to $10^{-8}$ mbar. The specifications of the system are as follows:

*Vacuum vessel*

Volume: 80 liters

Area of the walls on the high-vacuum side: 7065 cm$^2$

Specific outgassing rate of the walls after 1 h in high vacuum: $10^{-9}$ mbar s$^{-1}$ cm$^{-2}$

Permissible leak rate: $10^{-9}$ mbar s$^{-1}$

*Oil-diffusion pump*

Pumping speed for air: 300 liters s$^{-1}$ measured at $4 \times 10^{-3}$ mbar

Ultimate pressure: $4 \times 10^{-10}$ mbar (AN 175 fluid)

Connecting pipe between the oil-diffusion pump and the vessel

Length between the ports or the pump and the vessel: 12.0 cm

Inner diameter: 10.0 cm

*Temperature of the vessel*

$T = 23°$C

**Solution 10.11**

Since there are no elastomers, polymers, or glass parts in the system, the time to pump down the vessel is calculated from Eq. 10.14, where the second term is 0. Hence

$$t = Q_{\mathrm{Dm}} t'/[S_n(p_t - p_u - Q_W/S_n)]$$

Calculate $Q_{\mathrm{Dm}}$ from Eq. 10.15:

$$Q_{\mathrm{Dm}} = q_{\mathrm{Dm}} A$$

Substitute

$q_{\mathrm{Dm}} = 10^{-9}$ mbar liters s$^{-1}$ cm$^{-2}$
$A = 7065$ cm$^2$

Thus

$$Q_{\mathrm{Dm}} = 10^{-9} \text{ mbar liters s}^{-1}\text{ cm}^{-2} \times 7065 \text{ cm}^2$$

$$= 7.065 \times 10^{-6} \text{ mbar liters s}^{-1}$$

Calculate the conductance of the pipe that connects the pump to the vessel from Eqs. 6.18, 6.17a:

$$L' = L[1 + 1/(3 + 3L/7r)]$$

$$C_{\mathrm{mp}} = 3.64(T/M)^{1/2} A(8r/3L')$$

Note that the molecular conductance of the pipe $C_{\mathrm{mp}}$ is independent of any end effects and transmission probabilities. Substitute

$L = 12$ cm
$d = 10$ cm, $r = 5$ cm
$T = 273 + 23 = 296$ K
$M = 28.98$ g mol$^{-1}$ (see Table 1.2)
$A = \pi r^2 = 3.14 \times 5^2$ cm$^2 = 78.5$ cm$^2$

Hence

$$L' = 12 \text{ cm} \,[1 + 1/(3 + 3 \times 12 \text{ cm}/7 \times 5 \text{ cm})] = 14.98 \text{ cm}$$

$$C_{\mathrm{mp}} = 3.64(296/28.98)^{1/2} \times 78.5 \times [8 \times 5/(3 \times 14.98)]$$

$$= 812.82 \text{ liters s}^{-1}$$

Calculate the net (effective) pumping speed $S_n$ at the port of the vessel from

Eq. 4.41, written as

$$S_n = S_P C_{mp}/(S_p + C_{mp})$$

Substitute

$$S_P = 300 \text{ liters s}^{-1}$$
$$C_{mp} = 812.82 \text{ liters s}^{-1}$$

Hence

$$S_n = 300 \text{ liters s}^{-1} \times 812.82 \text{ liters s}^{-1}/(300 \text{ liters s}^{-1} + 812.82 \text{ liters s}^{-1})$$
$$= 219 \text{ liters s}^{-1}$$

Substitute in Eq. 10.14

$$Q_{Dm} = 7.065 \times 10^{-6} \text{ mbar liters s}^{-1}$$
$$t' = 3600 \text{ s (see Section 10.6.1)}$$
$$S_n = 219 \text{ liters s}^{-1}$$
$$p_t = 10^{-8} \text{ mbar (pressure in the vessel at time } t)$$
$$p_u = 4 \times 10^{-10} \text{ mbar (ultimate pressure of the pump)}$$
$$Q_w = 10^{-9} \text{ mbar liters s}^{-1} \text{ (permissible leak rate)}$$

The time required to evacuate the vessel is

$$t = 7.065 \times 10^{-6} \text{ mbar liters s}^{-1} \times 3600 \text{ s}/ \left[219 \text{ liters s}^{-1} \left(10^{-8} \text{ mbar} - 4 \times 10^{-10} \text{ mbar} - 10^{-9} \text{ mbar liters s}^{-1}/219 \text{ liters s}^{-1}\right)\right]$$
$$= 12098 \text{ s} = 3 \text{ h } 21 \text{ min.}$$

## References


Delafosse, J., and Mongodin, G. (1961), *Le Vide*, 92, 107.

Diels, K., and Jaeckel, R. (1958), *Letbold Vakuum Taschenbuch*, Springer-Verlag, Berlin.

Hablanian, M. R. (1979), in *Methods of Experimental Physics*, Vol. 14, *Vacuum Physics and Technology*, G. L. Weissler and R. W. Carlson (eds.), Academic Press, New York.

# 11

# Pumping Steady State

## *11.1 Pumping Steady State with Localized Gas Load*

### 11.1.1 Ultimate Pressure of a Vacuum System

The ultimate pressure $p_{us}$ of a vacuum system is determined by the gas load. Thus

$$p_{us} = Q_G/S \qquad \text{(e.g., Pa)} \tag{11.1}$$

where $Q_G$ (e.g., Pa $m^3$ $s^{-1}$) is the gas load (see Eq. 8.1) at almost constant rate and $S$ (e.g., $m^3$ $s^{-1}$) is the pumping speed of the pump. Two situations are possible: (a) $S = S_P$, where the pump is directly connected to the vessel; and (b) $S = S_n$, where the pump and the vessel are connected by vacuum components (pipe, baffle, etc.). In this case $S_n$ can be calculated from Eq. 4.41.

Equation 11.1 holds true when the pumping speed $S$ is constant within the operational range of pressures that the pump can produce.

## *11.2 Pumping Steady State with Distributed Gas Load*

The gas load at almost constant rate yielded by the walls on the high-vacuum side of a long pipe (e.g., storage rings) can be considered uniformly distributed along the length of the pipe (see Fig. 11.1). To reach a certain degree of low pressure, several pumps (diffusion, turbomolecular) are connected to the pipe at equal distances.

The pressures at the port of the pump and at different locations along the pipe, between two consecutive pumps (see Fig. 11.1) are as follows:

Pressure $p_p$ at the port of the pump:

$$p_P = q_{Dm} o L_p / S_P \qquad \text{(e.g., mbar)} \tag{11.2}$$

where $q_{Dm}$ (e.g., mbar liters $s^{-1}$ $cm^{-2}$) is the specific outgassing rate of the metallic surfaces; $o$ (e.g., cm), the perimeter of the pipe; $L_p$ (e.g., cm), the

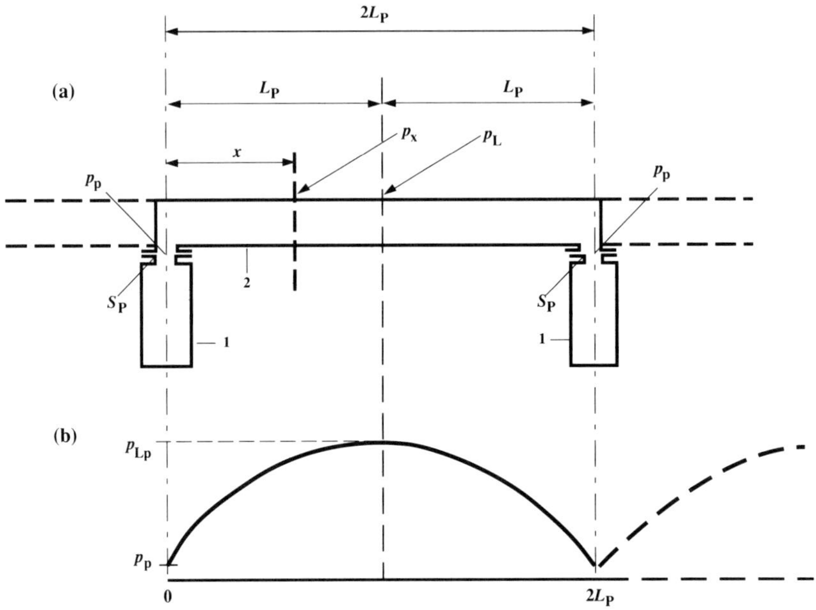


**Figure 11.1** Schematic of a pumping module used to maintain a required pressure in a long pipe: (a) 1—diffusion pump, 2—pipe of length $2L_p$, $S_P$—pumping speed at the port of the pump 1, $x$—distance along the pipe measured from the axis through the pump, $p_x$, $p_{Lp}$—pressures at different locations along the pipe; (b) pressure distribution along the pipe, measured between the axes through two successive pumps.

length of the pipe; and $S_P$ (e.g., liters $s^{-1}$), the pumping speed at the port of the pump.

Pressure $p_x$ at a distance $x$ from the port of the pump:

$$p_x = q_{Dm}o(L_p/S_P) + (x/C) - x^2/(2CL_p) \quad \text{(e.g., mbar)} \quad (11.3)$$

where C (e.g., liters $s^{-1}$) is the conductance of the pipe of length $L_p$; and $x$ (e.g., cm), the distance from the pump.

Pressure drop at a distance $x$ from the port of the pump:

$$p_x - p_P = q_{Dm}o[(x/C) - x/(2CL_p)] \quad \text{(e.g., mbar)} \quad (11.4)$$

Pressure drop at half distance between two consecutive pumps:

$$p_{Lp} - p_P = q_{Dm}oL_p/2C \quad \text{(e.g., mbar)} \quad (11.5)$$

Note that *the pressure drop* (Eqs. 11.4 and 11.5) *is independent of the pumping speed* $S_P$. *That means that pressure will not drop below the value predicted by these equations, regardless of the pumping speed of the pump.* The pressure distribution between two adjacent pumps is parabolic (see Fig. 11.1b).

For more details concerning the pressure drop in steady-state pumping, see Welch (1973), Calder (1974), Blechschmidt (1978), Miyahara (1986), Gottwald (1973), and Paul (1973).

## *Exercises*

**Exercise 11.1**

Select the pump required to maintain a pressure of $2 \times 10^{-1}$ mbar in a vacuum system in which a flow of air of $3.95\ \text{cm}^3\ \text{s}^{-1}$ at atmospheric pressure is expected. The volume of the pipe that connects the pump to the system is negligible compared to the volume of the system. Disregard the gas load due to outgassing.

**Solution 11.1**

Since the volume of the pipe that connects the pump to the vessel is negligible compared to the volume of the vessel, the pump and the vessel can be considered to be directly connected. Therefore, use Eq. 11.1 written as

$$S_P = Q_G/p_{us}$$

According to Eq. 8.1

$$Q_G = Q_D + Q_E + Q_L + Q_{Per}$$

Since $Q_D = 0$, $Q_E = 0$, $Q_{Per} = 0$, it follows that

$$Q_G = Q_L$$

$$Q_L = 1013\ \text{mbar} \times 3.95\ \text{cm}^3\ \text{s}^{-1}$$

$$= 1013\ \text{mbar} \times 3.95\ \text{cm}^3\ \text{s}^{-1} \times 10^{-3}\ \text{liters cm}^{-3} = 4.0\ \text{mbar liters s}^{-1}$$

Substitute

$$Q_G = Q_L = 4.0\ \text{mbar liters s}^{-1}$$
$$p_{us} = 2 \times 10^{-1}\ \text{mbar}$$

Thus

$$S_P = 4.0\ \text{mbar liters s}^{-1}/2 \times 10^{-1}\ \text{mbar} = 20\ \text{liters s}^{-1}$$

$$= 20\ \text{liters s}^{-1} \times 10^{-3}\ \text{m}^3\ \text{liter}^{-1} \times 3600\ \text{s h}^{-1} = 72\ \text{m}^3\ \text{h}^{-1}$$

According to Fig. 9.2, mechanical pumps are the most economical choice for handling this job. Allowing for a 20% pressure drop along the pipe connecting the pump to the vessel, and taking into account the commercial range of pumps available, one selects a 128-$m^3$ $h^{-1}$ double-stage rotary-vane vacuum pump with gas ballast.

## References

Blechschmidt, D. (1978), *J. Vac. Sci. Technol.*, *15*, 1175.
Calder, R. (1974), *Vacuum*, *24*, 437.
Gottwald, B. A. (1973), *Vakuum-Technik*, *22*, 106, 141.
Miyahara, Y. (1986), *J. Vac. Sci. Technol.* A4 *1*, 111.
Paul, G. (1973), *Vakuum-Technik*, *22*, 201.
Welch, K. M. (1973), *Vacuum*, *23*, 271.

# Appendix:

## System of Units, Physical Constants, Conversion Factors, and Relation among Temperature Scales

**Table A.1**
The International System of Units SI:[a] Base Units

| Quantity | Unit | Symbol |
|---|---|---|
| Length | meter | m |
| Mass | kilogram | kg |
| Time | second | s |
| Electric current | ampere | A |
| Thermodynamic temperature | kelvin[b] | K |
| Amount of substance | mole | mol |
| Luminous intensity | candela | cd |

[a]Adapted from the *International Standard ISO 1000*, 2nd ed., 1981-02-15.
[b]*Note:* The zero point of the Kelvin scale is $T_0 = 273.16$ K. If $t$ and $T$ represent the temperatures on the Celsius and Kelvin scales, respectively, then $t = T - T_0$.

**Table A.2**
The International System of Units SI:[a] Supplementary Units

| Quantity | Unit | Symbol |
|---|---|---|
| Plane angle | radian | rad |
| Solid angle | steradian | sr |

[a]Adapted from the *International Standard ISO 1000*, 2nd ed., 1981-02-15.

**Table A.3**
Some Derived Units of the SI System[a]

| Quantity | Name | Symbol |
|---|---|---|
| Area | square meter | $m^2$ |
| Volume | cubic meter | $m^3$ |
| Density (mass density) | kilograms per cubic meter | $kg\ m^{-3}$ |
| Speed velocity | meters per second | $m\ s^{-1}$ |
| Angular velocity | radians per second | $rad\ s^{-1}$ |
| Acceleration | meters per second squared | $m\ s^{-2}$ |
| Angular acceleration | radians per second squared | $rad\ s^{-2}$ |
| Force | newton | N |
| Pressure, stress | pascal | Pa |
| Viscosity, dynamic | See note[b] | $Pa\ s^{-1}$ |
| Viscosity, kinematic | See note[b] | $m^2\ s^{-1}$ |
| Work, energy | joule | J |
| Quantity of heat, power | watt | W |
| Quantity of electricity, electric charge | coulomb | C |
| Potential difference, electromotive force | volt | V |
| Electrical resistance | ohm | Ω |
| Frequency | hertz | Hz |
| Capacitance | farad | F |
| Magnetic flux | weber | Wb |
| Inductance | henry | H |
| Magnetic flux density, magnetic induction | tesla | T |
| Specific heat capacity | joules per kilogram and degree K | $J\ kg^{-1}\ K^{-1}$ |
| Thermal conductivity | watts per meter and degree K | $W\ m^{-1}\ K^{-1}$ |

[a]Adapted from the *International Standard ISO 1000*, 2nd ed., 1981-02-15.
[b]*Note:* The terms "poise" and "stokes" for dynamic and kinematic viscosity, respectively, belong to the CGS system of units.

**Table A.4**
Systems of Units: General Information

*Coherent system of units* A system of units in which quantities other than base units (see Table A.1) are derived by simple multiplication or division without use of numerical factors.
*Inch-pound units* Units based on the yard and the pound commonly used in English-speaking countries. The same unit may differ in magnitude in the United States and the United Kingdom (e.g., ounce fluid).
*Technical terms used with any system of units:*
*Digit* Any one of 10 Arabic-numeral symbols from 0 to 9.
*Figure* (numerical) The arithmetic value expressing one or more digits.
*Nominal value* A value assigned for the purpose of convenient designation, existing in name only.
*Significant digit* Any digit considered necessary to derive a value or a quantity.

**Table A.5**
Units Belonging to Other Systems Not Recommended by SI

| Quantity | Name of unit | Symbol | Expression in terms of base or supplementary SI units |
|---|---|---|---|
| Force | dyne | dyn | 1 dyn = $1.0 \times 10^{-5}$ N |
| Energy, work | erg | erg | 1 erg = $1.0 \times 10^{-7}$ J |
| Temperature | Rankine | °R | 1°R = 9/5 K |
| | Fahrenheit | °F | 1°F = 9/5 K − 460 |
| Quantity of heat | calorie | cal | 1 cal = 4.186 J |
| | British thermal unit | BTU | 1 BTU = $1.055 \times 10^3$ J |
| Pressure | physical atmosphere | atm | 1 atm = $1.01 \times 10^5$ Pa ($Nm^{-2}$)[a] |
| | technical atmosphere | at | 1 at = $9.81 \times 10^4$ Pa ($Nm^{-2}$)[a] |
| | Torr[b] | Torr | 1 Torr = 133.3 Pa ($Nm^{-2}$)[a] |
| | millimeter of mercury column | mm Hg | 1 mm Hg = 133.3 Pa ($Nm^{-2}$)[a] |
| | millimeter of water column | mm $H_2O$ | 1 mm $H_2O$ = 9.8 Pa ($Nm^{-2}$)[a] |
| | inch of | in. Hg | 1 in. Hg = 3386 Pa ($Nm^{-2}$)[a] |
| | mercury column | in. Hg | |
| | % vacuum | — | |

[a]For conversion factors exact within 6 digits after the decimal point, see Moss (1987).
[b]1 Torr differs from 1 mm Hg by 1.4 parts in $10^7$.

**Table A.6**
Some Fundamental Physical Constants[a]

| Quantity | Symbol | Value | Units |
|---|---|---|---|
| Avogadro's constant[b] | $N_A$ | 6.0221367 | $1.0 \times 10^{23}$ molec. $mol^{-1}$ |
| Boltzmann's constant, $RN_A^{-1}$ | $k$ | 1.380658 | $1.0 \times 10^{-23}$ J $K^{-1}$ |
| Stefan–Boltzmann's constant | | 5.67051 | $1.0 \times 10^{-8}$ W $m^{-2}$ $K^{-4}$ |
| Planck's constant | $h$ | 6.6260755 | $1.0 \times 10^{-34}$ J s |
| Molar gas constant (universal gas constant) | $R$ | 8.314510 | J $mol^{-1}$ $K^{-1}$ |
| Molar volume (ideal gas), $RTp^{-1}$ at 273.16 K and $1.01325 \times 10^5$ Pa | $V_M$ | 22,414.10 | $cm^3$ $mol^{-1}$ |
| Loschmidt's constant[b] | $n_0$ | 2.686763 | $1.0 \times 10^{25}$ molec. $m^{-3}$ |
| Elementary charge | $e$ | 1.60217733 | $1.0 \times 10^{-19}$ C |
| Electron rest mass | $m_e$ | 9.109389 | $1.0 \times 10^{-31}$ kg |
| Neutron rest mass | $m_n$ | 1.6749286 | $1.0 \times 10^{-27}$ kg |
| Proton rest mas | $m_p$ | 1.672623 | $1.0 \times 10^{-27}$ kg |
| Speed of light in vacuum | $c$ | 299,792,458 | m $s^{-1}$ |

[a]Adapted from Cohen and Taylor (1988, p. BG 11).
[b]Avogadro's and Loschmidt's "constants" are also known as Avogadro's and Loschmidt's "numbers."

**Table A.7**
Physical Constants of Standard Substances and Internationally Accepted Value of Acceleration Due to Gravity[a]

| Substance[b] | Density $\rho$ (kg $dm^{-3}$) | $T$ (°C) | $P$ (Pa) |
|---|---|---|---|
| Air free of carbon monoxide (a) | $1.2928 \times 10^{-3}$ | 0 | 101,325 |
| Mercury (b, c) | 13.59508 | 0 | 101,325 |
| Water, air-free (d) | 0.99987 | 3.98 | 101,325 |

[a]Acceleration due to gravity (*d*): $g = 9.80665$ m $s^{-2}$ = 32.174 05 ft $s^{-2}$.
[b]*Sources:* (a) *Landolt-Börnstein Physikalisch–Chemische Tabellen* (1967), (b) Cook (1961), (c) Cook and Stone (1957), (d) CRC (1986–1987).

**Table A.8**
Prefixes Used for Measurement[a]

| Prefix | Power of ten | Symbol |
|---|---|---|
| tera | $1.0 \times 10^{12}$ | T |
| giga | $1.0 \times 10^{9}$ | G |
| mega | $1.0 \times 10^{6}$ | M |
| kilo | $1.0 \times 10^{3}$ | k |
| mili | $1.0 \times 10^{-3}$ | m |
| micro | $1.0 \times 10^{-6}$ | $\mu$ |
| nano | $1.0 \times 10^{-9}$ | n |
| pico | $1.0 \times 10^{-12}$ | p |
| femto | $1.0 \times 10^{-15}$ | f |
| atto | $1.0 \times 10^{-18}$ | a |

[a]Adapted from *International Standard ISO 1000*, 2nd ed., 1981-02-15.

**Table A.9**
The Greek Alphabet

| | | |
|---|---|---|
| Alpha | A | $\alpha$ |
| Beta | B | $\beta$ |
| Gamma | $\Gamma$ | $\gamma$ |
| Delta | $\Delta$ | $\delta$ |
| Epsilon | E | $\epsilon$ |
| Zeta | Z | $\zeta$ |
| Eta | H | $\eta$ |
| Theta | $\Theta$ | $\theta$ |
| Iota | I | $\iota$ |
| Kappa | K | $\kappa$ |
| Lambda | $\Lambda$ | $\lambda$ |
| Mu | M | $\mu$ |
| Nu | N | $\nu$ |
| Xi | $\Xi$ | $\xi$ |
| Omicron | O | $o$ |
| Pi | $\Pi$ | $\pi$ |
| Rho | P | $\rho$ |
| Sigma | $\Sigma$ | $\sigma$ |
| Tau | T | $\tau$ |
| Upsilon | $\Upsilon$ | $\upsilon$ |
| Phi | $\Phi$ | $\phi$ |
| Chi | X | $\chi$ |
| Psi | $\Psi$ | $\psi$ |
| Omega | $\Omega$ | $\omega$ |

Capitals: $\Gamma\Delta\Theta\Lambda\Xi\Pi\Sigma\Phi\Psi\Omega$
Lowercase: $\alpha\beta\gamma\delta\epsilon\zeta\eta\theta\iota\kappa\lambda\mu\nu\xi o\pi\rho\sigma\tau$ $\upsilon\phi\chi\psi\omega$

**Table A.10**
Conversion Factors for Length

| | km | m | cm | mile | ft | in. |
|---|---|---|---|---|---|---|
| kilometer | 1.0 | $1.0 \times 10^3$ | $1.0 \times 10^5$ | $6.214 \times 10^{-1}$ | $3.281 \times 10^3$ | $3.937 \times 10^4$ |
| meter | $1.0 \times 10^{-3}$ | 1.0 | $1.0 \times 10^2$ | $6.214 \times 10^{-4}$ | 3.281 | $3.937 \times 10$ |
| centimeter | $1.0 \times 10^{-5}$ | $1.0 \times 10^{-2}$ | 1.0 | $6.214 \times 10^{-6}$ | $3.281 \times 10^{-2}$ | $3.937 \times 10^{-1}$ |
| mile | 1.609 | $1.609 \times 10^3$ | $1.609 \times 10^5$ | 1.0 | $5.280 \times 10^3$ | $6.336 \times 10^4$ |
| foot | $3.048 \times 10^{-4}$ | $3.048 \times 10^{-1}$ | $3.048 \times 10$ | $1.894 \times 10^{-4}$ | 1.0 | 12 |
| inch | $2.540 \times 10^{-5}$ | $2.540 \times 10^{-2}$ | 2.540 | $1.578 \times 10^{-5}$ | $8.333 \times 10^{-2}$ | 1.0 |

1 yard = 3 ft
1 rod = 16.5 ft
1 mil = $10^{-3}$ in.
1 Å = $10^{-10}$ m

**Table A.11**
Conversion Factors for Area

| | $m^2$ | $cm^2$ | $ft^2$ | $in.^2$ |
|---|---|---|---|---|
| square meter | 1.0 | $1.0 \times 10^4$ | $1.076 \times 10$ | $1.55 \times 10^3$ |
| square centimeter | $1.0 \times 10^{-4}$ | 1.0 | $1.076 \times 10^{-3}$ | $1.55 \times 10^{-1}$ |
| square foot | $9.29 \times 10^{-2}$ | $9.29 \times 10^2$ | 1.0 | $1.44 \times 10^2$ |
| square inch | $6.452 \times 10^{-4}$ | 6.452 | $6.944 \times 10^3$ | 1.0 |

**Table A.12**
Conversion Factors for Volume

| | $m^3$ | $cm^3$ | liter | $ft^3$ | $in.^3$ |
|---|---|---|---|---|---|
| cubic meter | 1.0 | $1.0 \times 10^6$ | $1.0 \times 10^3$ | $3.531 \times 10$ | $6.102 \times 10^4$ |
| cubic centimeter | $1.0 \times 10^{-6}$ | 1.0 | $1.0 \times 10^{-3}$ | $3.531 \times 10^{-5}$ | $6.102 \times 10^{-2}$ |
| liter | $1.0 \times 10^{-3}$ | $1.0 \times 10^3$ | 1.0 | $3.531 \times 10^{-2}$ | $6.102 \times 10$ |
| cubic foot | $2.832 \times 10^{-2}$ | $2.832 \times 10^4$ | $2.832 \times 10$ | 1.0 | $1.728 \times 10^3$ |
| cubic inch | $1.639 \times 10^{-5}$ | $1.639 \times 10$ | $1.639 \times 10^{-2}$ | $5.785 \times 10^{-4}$ | 1.0 |

1 U.S. fluid gallon = 4 U.S. fluid quarts
= 8 U.S. pints
= 128 U.S. fluid ounces
= 231 $in.^3$
1 British Imperial gallon = 277.4 $in.^3$

**Table A.13**
Conversion Factors for Mass

| | kg | g | slug |
|---|---|---|---|
| kilogram | 1.0 | $1.0 \times 10^3$ | $6.852 \times 10^{-2}$ |
| gram | $1.0 \times 10^{-3}$ | 1.0 | $6.852 \times 10^{-5}$ |
| slug | $1.459 \times 10$ | $1.459 \times 10^4$ | 1.0 |

**Table A.14**
Conversion Factors for (Mass) Density

| | kg m$^{-3}$ | g cm$^{-3}$ | slugs ft$^{-3}$ |
|---|---|---|---|
| kilograms per cubic meter | 1.0 | $1.0 \times 10^{-3}$ | $1.94 \times 10^{-3}$ |
| grams per cubic centimeter | $1.0 \times 10^3$ | 1.0 | 1.94 |
| slugs per cubic feet | $5.154 \times 10^2$ | $5.154 \times 10^{-1}$ | 1.0 |

**Table A.15**
Conversion Factors for Force

| | N | dyn | lb | pdl |
|---|---|---|---|---|
| Newton | 1.0 | $1.0 \times 10^5$ | $2.248 \times 10^{-1}$ | 7.233 |
| dyne | $1.0 \times 10^{-5}$ | 1.0 | $2.248 \times 10^{-6}$ | $7.233 \times 10^{-5}$ |
| pound | 4.448 | $4.448 \times 10^5$ | 1.0 | $3.217 \times 10$ |
| poundal | $1.383 \times 10^{-2}$ | $1.383 \times 10^4$ | $3.108 \times 10^{-2}$ | 1.0 |

**Table A.16**
Conversion Factors for Energy, Work, and Heat

| | J | erg | cal | BTU |
|---|---|---|---|---|
| J | 1.0 | $1.0 \times 10^{7}$ | $2.388 \times 10^{-1}$ | $9.481 \times 10^{-4}$ |
| erg | $1.0 \times 10^{-7}$ | 1.0 | $2.388 \times 10^{-8}$ | $9.481 \times 10^{-11}$ |
| cal | 4.187 | $4.187 \times 10^{7}$ | 1.0 | $3.969 \times 10^{-3}$ |
| BTU | $1.055 \times 10^{3}$ | $1.055 \times 10^{10}$ | $2.52 \times 10^{2}$ | 1.0 |
| ft lb | 1.356 | $1.356 \times 10^{7}$ | $3.238 \times 10^{-1}$ | $1.285 \times 10^{-3}$ |
| hp h | $2.685 \times 10^{6}$ | $2.685 \times 10^{13}$ | $6.413 \times 10^{5}$ | $2.545 \times 10^{3}$ |
| kW h | $3.6 \times 10^{6}$ | $3.6 \times 10^{13}$ | $8.598 \times 10^{5}$ | $3.413 \times 10^{3}$ |
| eV | $1.602 \times 10^{-19}$ | $1.602 \times 10^{-12}$ | $3.826 \times 10^{-20}$ | $1.519 \times 10^{-22}$ |
| | ft lb | hp h | kW h | eV |
| J | $7.376 \times 10^{-1}$ | $3.725 \times 10^{-7}$ | $2.778 \times 10^{-7}$ | $6.242 \times 10^{18}$ |
| erg | $7.376 \times 10^{-8}$ | $3.725 \times 10^{-14}$ | $2.778 \times 10^{-14}$ | $6.242 \times 10^{11}$ |
| cal | 3.088 | $1.560 \times 10^{-6}$ | $1.163 \times 10^{-6}$ | $2.614 \times 10^{19}$ |
| BTU | $7.779 \times 10^{2}$ | $3.929 \times 10^{-4}$ | $2.930 \times 10^{-4}$ | $6.585 \times 10^{21}$ |
| ft lb | 1.0 | $5.051 \times 10^{-7}$ | $3.766 \times 10^{-7}$ | $8.464 \times 10^{18}$ |
| hp h | $1.980 \times 10^{6}$ | 1.0 | $7.457 \times 10^{-1}$ | $1.676 \times 10^{23}$ |
| kW h | $2.655 \times 10^{6}$ | 1.341 | 1.0 | $2.247 \times 10^{25}$ |
| eV | $1.182 \times 10^{-19}$ | $5.967 \times 10^{-26}$ | $4.450 \times 10^{-26}$ | 1.0 |

**Table A.17**
Conversion Factors for Magnetic Flux

| | maxwell | weber |
|---|---|---|
| maxwell | 1.0 | $1.0 \times 10^{-8}$ |
| weber | $1.0 \times 10^{8}$ | 1.0 |

**Table A.18**
Conversion Factors for Magnetic Field

| | tesla[a] | gauss |
|---|---|---|
| tesla | 1.0 | $1.0 \times 10^{4}$ |
| gauss | $1.0 \times 10^{-4}$ | 1.0 |

[a] 1 tesla = 1 weber $m^{-2}$.

**Table A.19**
Fundamental Points of Thermodynamic Temperature Scale

| Name of scale | Symbol | Temperature value at | |
|---|---|---|---|
| | | Absolute zero | Triple point of water[a] |
| Kelvin | K | 0 | 273.16 |
| Celsius | °C | −273.15 | 0.01 |
| Fahrenheit | °F | −459.67 | 32.018 |
| Rankine | °R | 0 | 491.688 |

[a]The temperature of the triple point of water is the temperature at which the solid, liquid, and gaseous phases are in equilibrium.

**Table A.20**
Correlations between Temperature Scales

| Scale | Symbol | Steam point of water at 1 atm | Ice point of water at 1 atm | Absolute zero |
|---|---|---|---|---|
| Kelvin | K | 373 | 273 | 0 |
| Celsius | °C | 100 | 0 | −273 |
| Rankine | °R | 672 | 492 | 0 |
| Fahrenheit | °F | 212 | 32 | −460 |

0°F = 460°R
$T_C = T_K - 273$
$T_R = (9/5)T_K$
$T_F = T_R - 460 = (9/5)T_K - 460$
*Note:* The values of temperatures in this table have been rounded down to the nearest degree.

## References

Cohen, R. E., and Taylor, B. N. (1988), *Physics Today, 41* (8), Part 2, BG 13.
Conference Générale Des Poids et Mesures, *Comptes Rendus Des Séances De La 3e* (3rd) *Conference Générale des Poids et Mesures* (1901), Gauthier-Villars, Paris, p. 70.
Cook, A. H. (1961), *Philos. Trans. Roy. Soc., 254,* A1038, 125.
Cook, A. H., and Stone, W. B., (1957) *Philos. Trans. Roy. Soc., 250,* A978, 279.
CRC (1986–1987), *Handbook of Chemistry and Physics*, 67th ed., R. C. Weast (ed.), CRC Press, Boca Raton, Fla.
ISO (International Standards Organization), *1000 SI Units and Recommendations for the Use of Their Multiples and of Certain Other Units*, 1981-02-15.
*Landolt-Bornstein Physikalisch–Chemische Tabellen* (1967), 6th ed.,Volume IV, Section 4, Part a, Springer-Verlag, Berlin, p. 174.
Moss, T. A. (1987), *J. Vac. Sci. Technol.* A5, 5, 2962.

# Index